THE FABRIC OF THE HEAVENS

The Development of Astronomy and Dynamics

BY STEPHEN TOULMIN AND JUNE GOODFIELD:

The Architecture of Matter
The Discovery of Time

BY STEPHEN TOULMIN:

Cosmopolis
Human Understanding
The Place of Reason in Ethics
The Abuse of Casuistry (with Albert Jonsen)
Wittgenstein's Vienna (with Allan Janik)

BY JUNE GOODFIELD:

An Imagined World
Reflections on Science and the Media
The Siege of Cancer

THE FABRIC OF THE HEAVENS

The Development of Astronomy and Dynamics

Stephen Toulmin and June Goodfield

The University of Chicago Press
Chicago and London

The University of Chicago Press, Chicago 60637
The University of Chicago Press, Ltd., London

The University of Chicago Press Edition 1999
Printed in the United States of America
04 6 5 4 3 2

Library of Congress Cataloging-in-Publication Data

Toulmin, Stephen Edelston.
The fabric of the heavens : the development of astronomy and dynamics / Stephen Toulmin and June Goodfield.
p. cm.
Originally published: New York : Harper, 1962.
Includes bibliographical references and index.
ISBN 0-226-80848-3 (pbk.)
1. Cosmology—History. 2. Astronomy. I. Goodfield, June, 1927– . II. Title.
QB981.T8 1999
523.1—dc21
99-20400
CIP

∞ *The paper used in this publication meets the minimum requirements of the American National Standard for Information Sciences—Permanence of Paper for Printed Library Materials, ANSI Z39.48-1992.*

Contents

Acknowledgements 7
Authors' Foreword 11
GENERAL INTRODUCTION: COSMOLOGY 15
PART I: THE SOURCES OF THE OLD ORDER
1 CELESTIAL FORECASTING 23
The Sources 24
The Problems 26
The Background of the Problems 30
The Solution to the Problems 34
The Wider Issues 41
How the Babylonians Computed Conjunctions 48
2 THE INVENTION OF THEORY 52
The Sources 52
The Background 54
The Character of Greek Theory 58
The First Theories 64
From Ingredients to Axioms 69
Plato's Geometrical Astronomy 79
3 THE PREMATURE SYNTHESIS 90
Aristotle's Programme 91
Motion and Change 93
The Celestial Mechanism 105
The Size of the Earth's Sphere 112
4 DOUBTERS AND HERETICS 115
Patching up the Dynamics 117
Amending the Astronomy 119
Aristarchos' Heliocentric Theory 122

5 PHYSICS LOSES MOMENTUM 128
Four Questions 129
The Political Background to Late Greek Astronomy 131
The Scientific Background: The Retreat from Physics 133
The Scientific Background: An Acquisition 136
Ptolemy's Mathematical Astronomy 137
The Wider Revolt against Philosophy 145
Archimedes and the Circle 149

PART II: THE NEW PERSPECTIVE AND ITS CONSEQUENCES

6 THE INTERREGNUM 153
The Roundabout Journey 153
The Mediaeval Revival 158
The Background to Copernicus 161
Mediaeval Arguments about the Moving Earth 165
Copernicus: His Aim and his Theory 169
Copernicus: His Achievement 175

7 PREPARING THE GROUND 182
The Background of the New Science 182
The Work of Tycho Brahe 184
Galileo's Telescopic Discoveries 189
Johann Kepler's Astronomical Physics 198

8 THE CREATION OF MECHANICS 210
The Change from Aristotle to Newton 211
Treating Motion Mathematically 213
Motion and Force 221
The New Ideal: Straight-Line Motion 225

9 THE NEW PICTURE TAKES SHAPE 228
The Man and his Task 229
Newton's Argument 232
The Character of Newton's Achievement 238
The Unity of Craft and Theory 245

10 THE WIDENING HORIZON 250
The Loose Ends: (1) Planetary Inequalities 250
The Loose Ends: (2) The Mechanism of Gravity 256
The Larger-Scale Picture 261
The Wider Influences of Newton 264
Certainty and Scientific Theory 268

Index 273

A section of illustrations appears between pages 96 and 97.

Acknowledgements

For permission to reproduce halftone illustrations the authors are indebted to the following:

The Curators of the Bodleian Library for *An illumination from an Arabic text showing star-worship* (Plate 5); the Iraq Petroleum Co. for *The Ziggurat at Ur* (Plate 2); Percy Lund, Humphries & Co. Ltd. for *Cuneiform Tablet No. 101* from 'Astronomical Cuneiform Texts' by O. Neugebauer published for the Institute of Advanced Study, Princeton, New Jersey (Plate 1); the Mansell Collection for the *Bayeux Tapestry Comet* (Plate 7) and *Haymaking scene with astronomical data* from the 'Book of Hours' of the Duc de Berry (Plate 8); the Science Museum for *Armillary sphere*, British Crown Copyright (Plate 4), *Tycho Brahe's Mural Quadrant, photograph* (Plate 11), *Orrery*, British Crown Copyright (Plate 13), *Two Spiral Nebulas*, Photograph reproduced by permission of the Mount Wilson Observatory, California (Plate 14).

And for permission to reproduce line drawings:

The British Broadcasting Corporation for *A table showing the relation between the Lunar and Solar calendars* (page 34); the Clarendon Press for *Dante's scheme of the universe* from 'Studies in the History and Method of Science' by Charles Singer (page 163); Professor N. R. Hanson and Dr Harry Woolf for Figs. (pages 121, 141, 142) which appeared in an article on 'The Mathematical Power of Epicyclical Astronomy' in *Isis*, Seattle, June, 1960; Arthur Koestler for *Kepler's model of the universe* (page 200), and Van Nostrand & Co. Ltd. for *The retrograde motion of Mars* (page 27).

For permission to reprint copyright material the authors are indebted to the following:

Mr Lawrence Durrell and Messrs Faber & Faber Ltd for an extract from *Bitter Lemons*; Mr Siegfried Sassoon and Messrs Faber & Faber Ltd for the sonnet 'Grandeur of Ghosts' from his *Collected Poems*; and Professor Leon J. Richardson for his translation of The Halley Ode.

For permission to reprint translations from original sources the authors are indebted to:

The University of Chicago Press for an extract from *History of the Persian Empire* by A. T. Olmstead; Dover Publications for extracts from *Three Copernican Treatises* by Edward Rosen; Oxford University Press for extracts from *Mathematics in Aristotle* and *Aristarchos of Samos,* both by T. L. Heath; and Princeton University Press for the Table of Babylonian Computations from *The Exact Sciences in Antiquity* by O. Neugebauer.

In addition, the authors must acknowledge their debt to the following for translations which have been used in the course of preparing this book:

Early Greek Philosophy: J. Burnet (Messrs A. & C. Black Ltd); *Galileo's Dialogue concerning Two New Sciences:* Crew and De Salvio (Northwestern University Press); *Republic of Plato:* F. M. Cornford (Oxford University Press) and *Source Book in Greek Science:* M. R. Cohen and I. E. Drabkin (Harvard University Press).

When I have heard small talk about great men
I climb to bed; light my two candles; then
Consider what was said; and put aside
What Such-a-one remarked and Someone-else replied.

They have spoken lightly of my deathless friends
(Lamps for my gloom, hands guiding where I stumble),
Quoting, for shallow conversational ends,
What Shelley shrilled, what Blake once wildly muttered. . . .

How can they use such names and be not humble?
I have sat silent; angry at what they uttered.
The dead bequeathed them life; the dead have said
What these can only memorize and mumble.

Grandeur of Ghosts—*Siegfried Sassoon*

'. . . the rest
From man or angel the great Architect
Did wisely to conceal and not divulge
His secrets to be scann'd by them who ought
Rather admire; or, if they list to try
Conjecture, he his fabric of the heavens
Hath left to their disputes, perhaps to move
His laughter at their quaint opinions wide
Hereafter; when they come to model heaven
And calculate the stars, how they will wield
The mighty frame; how build, unbuild, contrive
To save appearances; how gird the sphere
With centric and eccentric scribbled o'er,
Cycle and epicycle, orb in orb.'

Paradise Lost, Book VIII—*John Milton*

Authors' Foreword

Those who rightly regret the contemporary division of educated men into 'two cultures'—or, at any rate, into two separate circles of conversationalists—can take comfort from one thing: namely, that we do still all grow up sharing a common conception of the world we live in, at any rate so far as concerns the main outlines of its structure and the chief phases of its development. This 'common-sense' view of the world, therefore, represents common ground shared by the two cultures, and any study of the process by which it came to its present form, and of its credentials, can hope to strike a chord in the minds of 'scientists' and 'humanists' alike. It was in this conviction that we started in 1957 at Leeds University a teaching course on *Origins of Modern Science*, which was followed with growing interest and equal success by students from both the Arts and Science Faculties; and we remain convinced that the evolution of scientific ideas—in particular, the embryology of our common-sense view of the world—is an important part of the region where scientific, historical, and literary studies overlap. On this common ground we may hope to restore the conversations between the two cultures which in earlier generations were taken for granted.

The present book is the first of four volumes, which will together form a connected series on *The Ancestry of Science*. The first three volumes comprise extended case-studies, centred on particular groups of topics which have played important parts in the evolution of our ideas. Here, in *The Fabric of the Heavens*, we shall be looking at the development of astronomy and dynamics, and the contribution these sciences have made to our cosmological picture. The second volume, on *The Architecture of Matter*, will concentrate on conceptions of material substance, both in physiology and in chemistry, and the gradual clarification of ideas about the special character of living things. The third volume will be a study of the way in which the historical dimension entered science: how the earlier, unhistorical vision of a static Nature was displaced by a developing one, and how this new historical approach has begun to spread from geology and zoology (where it bore its first fruit) into the physical sciences. In the final volume, we shall use the material gathered together in the earlier volumes in order to analyse the changing relations throughout history between science, literature, philosophy, technology, religion, and other aspects of human life.

Anyone who embarks on a work of historical synthesis and interpretation, such as this, inevitably places himself much in the debt of the scholars by whose devoted work he profits. In the present case, this debt is doubly worth acknowledging: for, during the last fifty years, the historical development of the natural sciences has been studied with a new care and disinterestedness, and all this work has begun to lead—especially since 1946—to an exciting new picture of the subject. Since little of this new picture has yet found a way into general literature, which tends rather to carry over into a more critical age the polemical prejudices of the nineteenth century, we have tried here to make the best use we could of the results of present-day scholarship. Our principal debts we have acknowledged in the reading-lists at the end of each chapter: however, we must include here a general expression of gratitude to all those whose work has led during the last few years to a better understanding of the scientific conceptions of our predecessors. Our gravest problem has been that of selection: in this, we have chosen to concentrate on a limited number of representative figures and to expound their views at some length, rather than make any pretence of being exhaustive.

'If you begin by treating the scientific ideas of earlier centuries as myths, you will end by treating your own scientific ideas as dogmas': we have tried, throughout this book, to display the developing character of the scientific endeavour, and to indicate why the different problems of cosmology came to be tackled in the order in which they did. If we are to understand even our own scientific ideas, and do more than simply manipulate with the most up-to-date calculi, we shall do well to study the strong points of the scientific systems which they displaced. From the quandaries and difficulties which delayed the formation of our modern 'common sense' we can discover best the true character and meaning of our twentieth-century conceptions.

STEPHEN TOULMIN
JUNE GOODFIELD

GENERAL INTRODUCTION:

Cosmology

THE task of these books is to illustrate and document the manner in which our chief scientific ideas have been formed. We shall begin with two sciences whose development has been very closely linked: astronomy and dynamics. These two sciences have tried to answer such questions as: What are the things one can see in the sky? How do they move? What makes them move? Are they at all like the things on the Earth around us, and do they move in the same way? About all these things, twentieth-century common sense has come to take the scientist's answers for granted, and our task is to follow out the sequence of steps by which our modern view was reached. How did the world look (we must ask) to the men who first tried to make sense of the things that happen in the sky above us? What conception did they have of the Sun and Moon, the stars and planets; and what problems had to be solved before we can recognize their point of view as our own, and say—as one does at the cinema—'This is where we came in'?

Common sense is a powerful mould. If we are to see the world through the eyes of the first astronomers, we must deliberately lay aside many beliefs and distinctions which nowadays we accept quite unthinkingly. For at the outset men faced the sky (as they did all aspects of Nature) in a state of far greater ignorance than we can easily imagine. They were confronted not by unanswered questions, but by problems as yet unformulated—by objects and happenings which had not yet been set in order, far less understood. (When you were a child, what would you yourself have made of the Heavens, with no adults to guide your eyes and thoughts?) To understand fully the scientific traditions which we have inherited, it is not enough to discover what our predecessors believed and leave it at that: we must try to see the world through their untutored eyes, recognize the problems which faced them, and so find out for ourselves why it was that their ideas were so different from our own.

What is required? Not just demoting beliefs which we now regard as established facts to the rank of daring speculations. Different situations gave rise in earlier times to different practical demands; different practical demands posed different intellectual problems; and the solution of these problems called for systems of ideas which in some respects are not even comparable with our own. Consider, for example, the question, How do the Earth and the Sun move relatively to one another? We are satisfied nowadays that the Earth goes round the Sun. We regard this as an established fact; though it was (and is) surely not *obvious*. What, then, could the first astronomers say about this? And what should we say about it ourselves, had we not grown up taking so much in the way of astronomical ideas for granted?

The correct answer is—that we should not even have understood the question. To lay aside our belief that the Earth goes round the Sun (our heliocentric *theory* of the planetary system) is to remove only the top layer of our astronomical garments. The men who first began to ask questions about the sky did not put forward a rival, geocentric theory, or indeed any theory at all—and why should they? Such questions as whether the Earth goes round the Sun or vice versa are comparatively sophisticated ones. The problems the first astronomers were concerned with were of a kind which did not give rise to theoretical questions, and it is a failure of our understanding if we discuss their ideas, as though they were committed to some theory—for instance, the geocentric theory. If anything, it was their whole *attitude*, and the whole range of practical problems they were directly concerned with, that were geocentric—focussed on their day-to-day lives, and so on this Earth, on which they and we alike have to find our way around and wrest a living.

So to see the world through their eyes we shall have to give up much more than our simple belief that the Earth goes round the Sun.

We must give up a whole habit of thought about the problems of planetary astronomy: for instance, the idea that the Sun and the Earth are similar sorts of heavenly bodies, and that our job is to work out their relative positions and speeds. Even to have got to the point of posing the choice between the geocentric and heliocentric theories meant acknowledging a similarity between the two bodies which could not be recognized all at once. The first astronomers, in fact, looked at the Heavens with quite different problems in mind, and so asked quite different questions.

This is only the beginning. Much else that we take for granted will have to be discarded. We assume nowadays, for instance, that the things which appear in the sky or come down out of it are of many different kinds. Stars are one sort of thing. Planets and comets are another. Meteors and the Polar auroras are a third. Clouds, thunderbolts, hailstones, and downpours of rain are different again; still more so, plagues of locusts. Yet we cannot expect to find our ancestors seeing that they all had to be explained in different terms. Stars and planets, comets and stars, the tails of comets and the Milky Way, meteors and thunderbolts, thunderbolts and earthquakes, thunderstorms and hailstorms, even torrents of hail and clouds of locusts: all these happenings and catastrophes will appear to the untheoretical observer more alike in both origins and effects than they do to us, for whom the basic scientific discoveries of 2000 years have become 'pure common sense'.

But what is common sense, anyway? One century's common sense is an earlier century's revolutionary discovery which has since been absorbed into the natural habits of thought. Of course, to begin with, everything to do with the sky was thought of in the same terms: it was the claim that these things were of many *different* kinds that had to be justified, in terms of discoveries which in 1000 B.C. had yet to be made. Where, in the sequence that stretches from stars at one end to locusts and earthquakes at the other, can one draw sharp lines? This was something which men found out only as they went along. The word 'meteorology' (the study of the-things-on-high) covered at first the science of *all* those things which happened above the Earth and so were inaccessible to close inspection. When we restrict the same word to the science of the weather, we preserve as a linguistic fossil an earlier phase in the evolution of ideas, when the geometrical scale of the universe was as yet unimagined, and when climatic and astronomical happenings fell naturally under the same heading. Our twentieth-century distinction between atmospheric happenings and astronomical ones understandably took some establishing.

If, at the outset, the classification of the heavenly objects inevitably remained confused, it is not surprising that the mutual relations and interactions of these objects remained obscure for even longer. Yet from the earliest times, certainly long before there was anything remotely resembling modern science, men were aware that many cycles and changes in the Heavens and on the Earth took place in step. As in spring the Sun's path climbed higher in the Heavens day by day, so vegetation began to flourish and crops to ripen: when in autumn it dropped back towards the horizon, vegetation, too, died back and the weather became colder. Similarly each month, as the Moon came to full, the tides of the sea rose higher and weak-witted men had fits of madness, being called 'lunatics': the same monthly cycle reappeared in the physiology of women. It was, therefore, a matter of plain observation that there were correspondences or harmonies between happenings in the Heavens—the 'macrocosm'—and happenings on the Earth and in man himself—the 'microcosm'.

A general belief in astrology was the natural result. Without the establishment of clear ideas of cause-and-effect, no one could confidently set a limit to these correspondences and harmonies. (They will reappear when we turn to the ancestries of chemistry and biology.) We are confident today that we know which objects in the Heavens can act directly upon terrestrial things, and by what means. For instance, the Moon by its gravitational pull produces the ebb and flow of the tides: other apparent harmonies, such as that between the lunar and menstrual cycles, we can no longer accept as instances of direct lunar action. But the boundary between astronomy and astrology inevitably remained blurred until men had worked out a satisfactory theory of planetary interactions, and this had to wait for Newton.

Turning from astronomy to the movement of bodies (the study

of dynamics) we must once more discard ideals and ideas whose correctness we normally take for granted. Since Galileo and Newton, dynamics has been made a branch of mathematics, and we no longer stop to question whether this can legitimately be done. In the complex equations of mathematical dynamics we assume that the way a body moves depends on how heavy it is, rather than (say) on its colour or its chemical make-up. Furthermore, dynamics—the science which explains why bodies change their position and motion as they do—is for us quite independent of other theories of change. Suppose some object we are studying alters its position, its shape, its material make-up, its colour, or its vital activities. We assume that each of these alterations is of a distinct kind, and we look to specialists in different sciences to explain them. We no longer see any but the most far-fetched analogies between (for example) a motor-car accelerating, a man growing old, a pool of water evaporating, and a dye fading. Dynamics, physiology, physical chemistry, and the rest, all have their own distinct subject-matters, explaining the processes which are their concern in terms of quite different sets of principles. We have given up the ambition to produce in science a general and comprehensive theory to explain all sorts of change.

Yet the distinctions between our present-day sciences are quite recent. If, in the beginning, men hoped to build up a general theory which would embrace all kinds of change, this ambition was from the point of view of method perfectly reputable. It has always been very much part of the scientific quest to devise theories covering as wide a range of happenings as our experience of things permitted. If today we draw lines beyond which, in our opinion, it is unprofitable to generalize further, that is because experience has forced us to do so; discovering these limits was an important stage in the evolution of our contemporary ideas. Even within the theory of 'motion', in the strict sense of the term, the boundaries have been shifting: different centuries have been preoccupied with quite different problems. Must we find causes to explain every movement—every change of position, that is? Or rather, should we look for causes only when moving bodies change their speed or direction? Even at this basic level, different centuries speak with different voices. For example, the belief that a dynamical theory should consist of mathematical equations, relating quantities such as momentum and force, is itself scarcely more than 400 years old. It has been universally accepted among physicists for only 250 years, and in earlier times was not even thought of as an ideal.

One last point must be mentioned. From the very beginning of science at least three strands have been present. Men have striven to understand the phenomena of Nature for several reasons—partly from sheer intellectual curiosity, partly to gain technical command over natural processes, but partly also from motives which were in effect religious. If the boundary between science and technology has always been open and fluctuating, so, too, has the boundary between science and religion. Men have tried to make sense of Nature, not just for the sake of the understanding itself or for the devices and skills that it has made possible, but also because they were concerned to find in the world around them objects worthy of their veneration—aspects of Nature to which they could justly give a religious response. The Earth was there to be ploughed, the Sky to be watched for signs. But the Earth and the Sky were also Divine agents and bore in themselves traces of the Divine handiwork.

Astrology, divination, the prediction of eclipses, the recognition of the Divine will, the timely control of ritual—all at first went hand in hand, depending equally upon one and the same kind of understanding. Alongside the geocentric *theory* of mediaeval astronomy, and the geocentric *attitude* of practical men for whom the sky was a perpetual calendar and guide, we shall find the geocentric *hierarchy*. In this religious picture, the relative position of the Earth and the Heavens symbolized and reflected their relative status in the Divine Scheme. The astronomical revolution of the sixteenth and seventeenth centuries undoubtedly gave rise to religious controversy; but it did so neither through an accident of history nor because of clerical obstinacy alone. The independence of astronomy and theology had not yet been clearly established. So, in order to win general acceptance for a heliocentric planetary theory, men had to overcome both scientific and religious objections—not only the scientific case for a geocentric theory, but also the religious predisposition in favour of a geocentric focus. This striking change within science inevitably had its impact outside. Accordingly, the evolution of our astronomical ideas shows particularly clearly how the rival demands of technology, theoretical understanding, and religious sentiment act and react upon each other.

Galileo, for instance, was directly concerned with certain restricted problems in dynamics and planetary theory. His demonstrations with inclined planes showed the superiority of his definition of acceleration; the things he saw through his telescope provided fresh arguments in favour of the Copernican system. But the *indirect* effects of his discoveries were equally profound: one can already see in the poetry of the seventeenth century the dramatic influence he was to have on our whole vision of the world. It has been the same with Newton's discovery of the law of gravitation. Edmund Halley, who persuaded Newton to publish the theories set out in the famous *Principia*, wrote a preface to that work in Latin hexameters. In this poem Halley did not attempt to put the law of gravitation into verse: rather, he emphasized the influence that Newton's new ideas were bound to have on other sides of human existence. Planetary theory was an important part of mathematics, certainly, but it was more than that—it had implications for all of us. Halley was surely right to recognize one important result of Newton's success: namely, the blow which he was striking against astrology. Once we understand (for instance) why comets move as they do, and reappear when they do, we need no longer treat them as portents:

> No longer need we fear
> When bearded stars appear.

The triumphs of the pure scientific intellect have, in this way, increased also Man's command over his own emotions and attitudes.

But this was the end of a long story. In the analysis that follows we shall dissect, logically and historically, the developing body of ideas from which our contemporary picture of the world has descended, working up through the successive layers in which these ideas were formed. The scope and problems of physical astronomy and planetary dynamics were not self-evident: nor could they become clear all at once. It was one task to predict the appearances of the heavenly bodies—their risings and settings, their eclipses, their speedings-up, slowings-down, and stationary points. It was another task to speculate about the physical character of the happenings we see taking place in the Heavens. Working out a plausible model of the planetary system was another problem again. And all these problems could be tackled perfectly well, without men's solving the problem which was eventually to prove crucial: the question, that is, of what *forces* act between the heavenly bodies. Our story begins before 1000 B.C., with men's first bare records of celestial phenomena and portents: it ends with the comprehensive theories of twentieth-century physical astronomy. Looking back at this sequence, we shall find different problems under discussion at each stage. The first task is to understand the developing character of these intellectual problems. Only if we take the trouble to see clearly what questions preoccupied astronomers in any century—and why—can we hope to be fair judges of the answers they found convincing.

Part I: The Sources of the Old Order

I

Celestial Forecasting

The shopping centres would be deserted for half a day after an incident; and then the people would slowly creep out again, wistfully breathing in the silent air, like animals sniffing the wind; and reassured, they would start to go about the hundred trivial tasks of the day which the automatism of ordinary life had made endearing, comprehensible—containing no element of prediction. . . .

One has seen rabbits scatter like this at the first report of a gun, only to re-emerge after half an hour and timidly come out to grass again—unaware that the hunter is still there, still watching. Civilians have no memory. Each new event comes to them on a fresh wave of time, pristine and newly delivered, with all its wonder and horror brimming with novelty. Only in dull offices with electric light burning by day the seekers sat, doggedly listing events in order to study their pattern, to relate past and present, so that like stargazers they might peer a little way into the darkened future.

SO LAWRENCE DURRELL, in his book *Bitter Lemons*, describes the impact on men's lives of the violence in Cyprus; and, in so doing, he incidentally describes also one of the starting-points of all science.

Before men can build a successful theory in any field of science they must be clear what it is that requires explanation. In due course, the theory (they hope) will make sense of the regularities observed in the things happening around us. But for a start the regularities themselves have to be recognized and analysed. The first step is to see that, over a period of time, events of some particular kind regularly recur: only then can we ask—Why? Until we have achieved a fair grasp of the nature of these regularities, we cannot hope to judge fairly between the theories put forward to account for them. Strikingly enough, in the field of astronomy, these two aspects of science—the careful study of regularities, and theory-building—had their origins in different places; and up to a certain point they developed independently.

Let us begin, then, by looking at a stage of astronomy that was (so far as we can tell) entirely pre-theoretical—the computational astronomy of the Babylonians. A century ago, we knew nothing about it at all.

THE SOURCES

Until recently our knowledge of the ancient empires of the Middle East was fragmentary, indirect, and often inaccurate. The writings of the classical Greeks reached modern Europe in a state of some disrepair, often after much copying, re-copying, and even multiple translations. These writings contained scattered references to the astrological traditions of the 'Chaldeans'. Some authors went so far as to admit that Mesopotamia and Egypt had been the home of intellectual traditions many centuries old, as well as of highly organized states. There were even suggestions that the early Greeks themselves, particularly those shadowy figures Thales and Pythagoras, had learned some of what they knew in Egypt or in Babylon. But it was impossible to reconstruct the actual content of the older traditions from these references alone, and to most Europeans the name of Babylon meant scarcely more than could be gleaned from the story of the Tower of Babel.

Men were, in consequence, little prepared for what archaeology and scholarship were to bring to light. During the last seventy years, our knowledge of Babylonia and its astronomy has been entirely transformed, and the process is still going on. Greek literature, history, and philosophy are known to us in most cases from copies made centuries after their original composition. But we are now in a position the historian can in many cases only dream of: we can handle and decipher the very documents on which Babylonian astronomers recorded their observations, predictions, and methods of computation centuries before Christ.

The archives of the great cities of Mesopotamia were kept on baked clay tablets which have preserved legibly to this day a mass of records whose very existence had been quite unsuspected. The inscriptions are in 'cuneiform' script, arrangements of wedge-shaped impressions made in soft clay with the edges and corners of a small rectangular stick and then put in the sun to bake hard. To begin with, most of the tablets found in the ruined cities of Ur, Babylon, and Uruk referred to matters of law or ritual, or recorded contracts, or gave inventories of animals or slaves or crops. Then one group of tablets turned up written in long columns, and headed with the names of Gods—which were also names of heavenly bodies. Deciphering these tablets has called for extreme ingenuity, but it has eventually become clear that they correspond very closely to the records of our own Nautical Almanac Office. (Plate 1 shows one of these tablets, and a transcription into our notation, together with an excerpt from the *Ephemeris* for 1960 for comparison.)

These tablets comprise planetary observations, tables predicting the motions and eclipses of the Moon, 'procedure-texts' setting out the arithmetical steps to be taken when calculating 'ephemerides' (daily positions of the planets), and a mass of similar material. There are also horoscopes: concerned at first with affairs of state, though we do find horoscopes being prepared for individuals as well from about 400 B.C. In addition, there are tablets recording the times and dates of earthquakes, plagues of locusts, and other natural disasters: these were prepared, one imagines, in the hope that some recurrent cycle could be detected in them, like the regular recurrences in astronomical events—for, if it could, the Diviners should then be able to forecast them, in the way they learned to do for lunar eclipses.

The picture we can build up from these relics is still incomplete, but it already shows that the respect the Classical Greeks expressed for the Babylonians was based upon more than mere legend. In two things, particularly, the Babylonian astronomers were the masters of us all: they kept continuous, dated records of celestial events from at least 747 B.C., and their best mathematical techniques were not excelled until quite recently. They maintained these skills without interruption even after Alexander the Great captured their cities in 331 B.C.: like any real scientific tradition, theirs was a continuing and developing one, whose most sophisticated achievements came as late as 150 B.C., in the Seleucid period.

In the oldest tablets, mathematical texts are quite separate from astronomical ones. Multiplication tables appear as early as 2000 B.C. A little later on, we find simple geometrical problems, which were solved in arithmetical or algebraic ways, quite unlike the more familiar methods developed by the Greek geometers (such as Euclid) 1500 years later. About the same time—around 1800 B.C.—the first star-catalogues were compiled, and planetary movements began to be recorded. To start with, all celestial positions were recorded by reference to recognizable constellations—reading, for example: 'three fingers-breadth from the tail of the Great Bear'. Only from around 450 B.C. do we find men locating astronomical events by using a precise numerical system of angular 'degrees': from then on positions are referred to the signs of the Zodiac. On this system the sky is arbitrarily divided into twelve zones of 30°, each zone having a particular constellation as its main feature. 'Angular arithmetic' now became possible: distances across the heavens could be added and subtracted just like distances on land, and highly accurate arithmetical techniques could be introduced for predicting celestial happenings.

The numerals used by the Babylonians for arithmetical calculations ran from 1 to 60, 61 to 3600, and so on; whereas our own 'decimal' numerals run from 1 to 10, 11 to 100, and so on. So naturally they employed these 'sexagesimal' numerals for recording angular degrees also. And in fact, we have taken over their system of angular measure unchanged and still employ it: where we now write an angle as 29°, 31′, 5″ (speaking of degrees, minutes, and seconds), they would simply write 29,31,5 as the successive digits of a sexagesimal fraction—indicating the number $29 + \frac{31}{60} + \frac{5}{(60)^2}$ as we do the decimal fraction 5·48, i.e. $5 + \frac{4}{10} + \frac{8}{(10)^2}$.

THE PROBLEMS

As far back as we have any record, men have always had reasons for studying the sky and the things going on in it. Settled agricultural life demands a practical grasp of the cycle of the seasons, so, although man's focus was on the earth where he lived, his attention was early on drawn to the comings and goings of the heavenly bodies: their appearances ('phenomena', as the Greeks called them), their disappearances, phases, and eclipses—those recurrent cycles whose existence is already obvious even before one has any theory or model of the Heavens. For centuries, indeed, the Babylonians studied this pageant with great sophistication, yet without—so far as we can tell—asking, as a matter of physics, what the objects in the celestial procession really were. They observed, recorded, and mastered the astronomical cycles for practical reasons, and seem to have ignored most questions of a speculative kind.

The basic problem which the Babylonians tackled—and largely solved—was the problem of celestial forecasting. Some things are easy enough to predict: as the seasons change, the constellations move gradually and uniformly across the sky, and reappear year by year in the same regular succession. But the Sun, the Moon, and the planets all move across the constellations at varying speeds, and in a pattern which is sometimes intricate.

Two examples will illustrate this. Consider first the planets: as the Babylonians knew, these do *not* move across the fixed stars in straight paths at constant speed. If you watch a planet such as Mars, plotting its position in the constellations every night as the months go by, the extreme irregularity of its path may well become

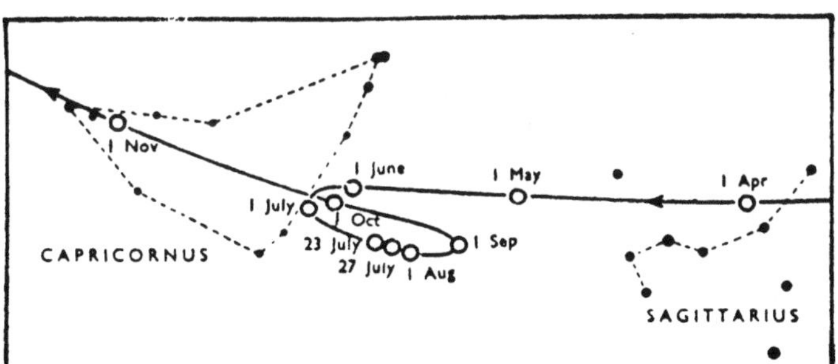

The retrograde motion of Mars in 1939

apparent. After moving steadily across the stars for some time, the planet periodically stops for a few days, then moves back for a few weeks, then stops once more, before finally resuming its normal eastward march. Join up the observed points and you will see that the planet has traced out a loop: as we now say, its motion has been 'retrograde'. (The Babylonians recorded and knew how to predict this retrograde motion of the planets.)

Even the Sun's visible motion is a complicated matter to analyse. One can again plot its position relative to the constellations, by noting the stars which appear in the western sky shortly after the sunset. Record the daily positions of the Sun in this way for a whole year, and join the points up on a star-map: the resulting smooth track is known as the 'ecliptic', and is the same every year. (The Moon and the planets also move across the stars roughly along the same line.) Looked at solely from the Earth, however, the Sun has a double motion: it not only moves along the the ecliptic but shares the daily and annual movement of the stars and the whole firmament—swinging right across the sky once every twenty-four hours, and changing its elevation with the seasons. This double nature of the Sun's visible movement must be kept clearly in mind, especially when we come to consider the problems which faced the Greeks. Judged against the horizon, for instance, the Sun appears to move westwards across the sky slightly faster than the Moon, which lags visibly behind it from one day to the next by about 12°. You will notice this most clearly around sunset for the few days just after New Moon: the Moon gets progressively higher in the sky at sunset, and sets correspondingly later, as the month goes on. Yet, considered relative to the background of the fixed stars, the Moon is actually moving along the ecliptic in an easterly direction *thirteen times as fast* as the Sun, covering in a month a distance which it takes the Sun a whole year to cover.

All this the Babylonians knew well. Perhaps the clearest account of the matter is given by the Roman author, Vitruvius, who included in his treatise on architecture instructions for the design and manufacture of sundials. These he prefaced with an account of the system of the Heavens which is probably taken directly from Demokritos, and large parts of it are believed by some authorities to describe the Babylonian view of the cosmos. In the passage which follows, speculations of obviously Greek origin are omitted, and the result is a reasonable summary of the way the Heavens must have appeared to the first systematic observers.

The system of Nature is so arranged that its upper Pole is high above the earth on the Northern side, while to the South the opposite Pole is hidden by the lower part of the Earth. In addition there is an inclined zone across the middle, which dips to the South of the Equator, and is figured with the twelve signs of the Zodiac. The twelve Zodiacal constellations roll around the Earth and the Sea along with the remaining stars, completing their path with every revolution of the heavens.

Now at the same time that these twelve signs, each of which occupies a twelfth part of the firmament, are turning continually from East to West, there wander across the firmament from West to East in a contrary direction, through the same signs, the Moon, Mercury, Venus, the Sun himself, as well as Mars, Jupiter and Saturn—just as though each was carried on its own step and had its own track. The Moon takes 28 days and about an hour more to complete its track from one side of the Zodiac back round to the same side, so completing the Lunar month.

The Sun, on the other hand, takes a month to cross a single sign: i.e. 1/12th part of the heavens. In this way he crosses all twelve signs in the course of twelve months, returning to the sign from which he started and so completing the space of one year. Thus the circle which the Moon runs round thirteen times in a year is measured out by the Sun only once in the same period.

The planets Mercury and Venus give the appearance of encircling the rays of the Sun, and so from time to time retreat backwards and slow up, lingering in their passage across the constellations when their tracks cross the ecliptic. This can be most clearly seen in the case of Venus which, when it is following the Sun, can be seen in the sky after sunset and on account of its bright light is called the Evening Star; whereas at other times it is ahead of the Sun and rises before the dawn, being then called Lucifer, the Bringer of Light. So these planets sometimes stay in one sign of the Zodiac for several days longer, and at other times enter the next sign more quickly than usual. Since they do not spend the same fixed number of days in each of the signs, they make up at one time by their speed for the amounts which they previously delayed; quickly making up their proper course after each delay.

The planet Mercury so travels across the firmament that it returns to the sign from which it began on the 360th day after starting its course. On an average it spends about 30 days in each sign of the Zodiac. At its fastest the planet Venus will cross one sign of the Zodiac in 30 days. But if it spends less than 40 days in any particular sign, it makes up the difference by delaying in one sign when it comes to its stationary point. So it completes its whole circuit across the firmament on the 485th day, reaching that sign from which it originally started on its journey.

Mars crosses the constellations and gets back to the beginning of its track on about the 683rd day: after going more quickly across some signs, it makes up for this when it reaches its stationary points. Jupiter, climbing more gently against the rotation of the heavens, takes about 360 days in each sign and comes back after eleven years 313 days to the sign in which it had been nearly twelve years before. Saturn, crossing each sign in 29 months and a few days more, takes 29 years 160 days to regain the sign in which it had been nearly thirty years previously.

From time to time, especially when they are in the same third of the sky as the Sun, these planets stop moving forwards and go back along tracks until the Sun has passed into the next sign of the Zodiac.

Notice that in all this we are concerned only with the positions of the Heavenly bodies, *as we see them*. No question of explanation has yet arisen. Of course, we can nowadays account for all these visible different motions—ecliptic, equatorial, and so on—by referring to the Earth's rotation in its axis, the Moon's path round the Earth, and the like. But for calculating where the Moon (say) will be on such-and-such an evening at sunset, these explanations are unnecessary.

The ideal the Babylonians aimed at was that of making all these motions as predictable as those of the stars. They were particularly concerned with two problems: (i) the problem of foretelling lunar and solar eclipses, (ii) the problem of calculating on which evenings the New Moon would first be visible on the western horizon.

THE BACKGROUND OF THE PROBLEMS

Why were the Babylonian astronomers so interested in these particular questions? The reasons throw light on the practical aims which Babylonian astronomy was expected to meet: divination, and the control of the calendar. On the one hand, celestial bodies were regarded as gods and all the major planets were given the names of gods. Their motions were thought to influence, not just the weather and the tides, but also the health and fortunes of men and of states. So the art of prophesying astronomical events was valued as yielding possible clues to the future welfare of the State. (Other forms of divination were used also: for instance, animals were sacrificed and their livers were examined for 'favourable' or 'unfavourable' signs.) Irrelevant though this reason may be to astronomy as we conceive it nowadays, it did make a careful study of the motions of the Heavenly bodies important to the Babylonians; and in the time of the great king and law-giver Hammurabi, about 1800 B.C., star-catalogues and rough planetary records were already being prepared, presumably under official auspices.

But though eclipses were interesting mainly as omens, there was a further compelling reason for the State authorities to encourage the pursuit of astronomy. This was the need, both civil and religious, for a uniform and reliable calendar. Though our own calendar, with its leap-years and its months of unequal lengths, may in some ways be confusing, at any rate it works out well enough for us not to have to think about it very much; especially as it is effectively universal, being used in Britain and New Zealand, Siberia and South America alike. In the ancient world, by contrast, many places had different calendar-systems, so that it could be a hard task to find out what date on the system accepted in Egypt (say) corresponded to a given date on another system, such as the Athenian one. Some of these calendars were based on the annual motion of the Sun, others on the phases of the Moon. The resulting difficulties were tolerable in a small city-state, and throughout the classical period in Athens the beginning of the new month was proclaimed by the local 'archons' (magistrates) when they saw the New Moon for themselves. But Mesopotamia was the heart of a large, centrally-directed land-empire. Commercial and official business alike called for a more predictable and uniform calendar, free of local variations and other irregularities. So it is not surprising to find that, along with his other reforms, Hammurabi ordered that a common calendar should be set up, to operate throughout his empire.

It is however one thing to issue a decree, another thing to put it into effect. The technical problems involved in setting up a uniform system of dates from scratch are greater than we nowadays realize. For short periods of time, the day is a perfectly satisfactory unit; but for anything of a more than day-to-day concern one needs a longer unit, and the obvious one to try first is the month—that is, the period from one New Moon to the next. In most of the countries of the ancient world, the word 'month' meant the period beginning at the *sunset* when the New Moon was first visible and ending on the next such occasion. This is still the official definition of the month in the religious calendars of the Jewish and Islamic worlds, and it is the business of the priest or Imam to declare the beginning of the month when he sees with his own eyes the crescent of the New Moon on the western horizon. The month begins at sunset and each successive sunset marks the start of a fresh day, until the cycle of lunar phases is complete.

Using this lunar month as a unit for dating runs one into two serious difficulties, and it required great ingenuity on the part of astronomers to overcome these difficulties, or mitigate them. The first practical problem arises because the number of days from one New Moon to the next is not always the same. Sometimes it is twenty-nine, sometimes it is thirty; the numerical average being slightly more than twenty-nine and a half. Further, the speeds of the Sun and Moon across the sky are not quite constant; and many other factors affect the actual visibility of the New Moon. For example: if the Moon is too close to the Sun at sunset, the daylight will still be too bright for the thin arc of the Moon to be seen. The angle between the ecliptic and the horizon also affects the visibility: in the summer, the Sun and Moon set more nearly at right angles to the horizon, but in the winter the angle is less and the distance between the two bodies may have to be greater before the Moon can be seen.

So it is a complicated problem to work out beforehand the point of first visibility, and whether any particular month is going to be twenty-nine or thirty days in length. The Athenians of the Golden

Age certainly never solved this, and the first effective method of predicting the new month well in advance was developed even in Babylon only around 300 B.C.: it represents one of the most sophisticated applications of Babylonian arithmetical techniques.

Another serious problem arises as soon as one tries to keep a calendar based on lunar months in step with the annual cycle of the seasons—that is, with the movements of the Sun. A year made up of twelve lunar months is eleven days too short: i.e. the Sun has eleven days to travel to complete its cycle. On this system, the new year will come sooner and sooner each time round, as compared (say) with the shortest and longest days, i.e. the solstices. On the other hand, a year consisting of thirteen lunar months will contain about nineteen days too many, and each new year will fall nearly three weeks later—judged by the Sun—than the one before.

There are various ways of getting round this difficulty. The Egyptians settled for a purely solar calendar, as we do ourselves. The actual *phases* of the Moon were disregarded, and the month was fixed arbitrarily to be thirty days long, irrespective of what the Moon looked like. Twelve months of thirty days together with five extra 'leap-days' at the end of the year made up a total of 365 days. Most other peoples, however, followed the Babylonians in taking a sequence of twelve lunar months as the basis of the year: the remaining eleven days were made up by inserting an extra month—'intercalating' it—whenever the solstices began to fall too far behind the legal calendar. Alternating occasional years of thirteen lunar months with the regular years of twelve prevented the calendar year from getting too far out of step with the Sun. Some degree of regularity can be introduced into such a mixed 'luni-solar' calendar, once it is discovered that nineteen solar years only differ from 235 lunar months by approximately half a day: one can then, in theory, add extra months in a regular cycle nineteen years long, and so keep the calendar in step with the seasons over a long period. And, in fact, we do find the cycle of intercalations becoming somewhat more regular from about 750 B.C. on, though complete command of the calendrical cycles was not achieved until 300 B.C. or later. Yet, even without this final step, the Babylonian calendar-system was so well administered and their records were so well kept that, from 626 B.C. up to A.D. 45, we can place all events dated on the Babylonian and Persian system with an uncertainty of one day at most. This accuracy we cannot approach in the case of classical Athens.

580	Apr. 4	575	Apr. 12	570	Apr. 16	565	Mar. 22
579	Mar. 26*	574	Apr. 1†	569	Apr. 4*	564	Mar. 11†
578	Apr. 14	573	Apr. 19	568	Apr. 23	563	Mar. 29*
577	Apr. 3*	572	Apr. 8*	567	Apr. 13	562	Apr. 17
576	Apr. 22	571	Apr. 26	566	Apr. 2		

A table showing the relation between the Lunar and Solar calendars

The years 580–562 B.C. were the last nineteen years of the reign of Nebuchadnezzar II. The beginning of the lunar year varied in this time between 11 March and 26 April, on our reckoning. Extra lunar months were added to seven of the years, in the spring in five cases (*), and in the autumn in the other two (†). (Nineteen solar years, of twelve calendar months each, correspond closely to 235 lunar months.)

THE SOLUTION TO THE PROBLEMS

How did the Babylonians set about solving these problems? And how did their discoveries, and their mathematical techniques, affect their ideas about the heavenly bodies? What sorts of things did they think the planets were, and how did they *explain* the observed heavenly motions?

They became skilled celestial forecasters, and in all that concerns the visible motions of the heavenly bodies they were extremely knowledgeable. Yet their methods were purely computational—complicated arithmetical sums carried out by following ingenious rules. So far as we can tell, they never went beyond the problems from which they began, or looked at them in a theoretical frame of mind. Their queries were all of the form: When this year will the constellation Orion first be visible in the sky? When shall we next see an eclipse of the Moon? When will Mars next halt in its tracks, and begin its retrograde motion? These were the questions they asked:

the methods of thought they employed enabled them to answer precisely these questions—*and these alone*.

What they did was to analyse the records of the heavenly motions in an arithmetical way. We can understand what this involved by considering how a similar sort of problem is solved these days: namely, how the tide-tables are prepared. For though, as a matter of theory, we understand in general terms why the sea rises and falls as it does, the task of *predicting* the times and heights of tides at a particular place is far too complicated to be worked out from first principles. In consequence, tide-tables are computed by arithmetical analysis of a sophisticated sort, developed empirically, by trial and error; and these methods are not dependent on any appeal to the theory of gravitation. This is not to say that the tides actually violate the laws of dynamics and gravitation: no one supposes that. But the task of applying these laws to predict tidal movements is too complex to be manageable, and the job can be done quite adequately by numerical analysis of the past records.

How do we do this? There is, of course, no difficulty in keeping records of the times and heights of tides at a given place, provided one is content with an accuracy of a few inches and a few minutes. Having got an extensive sequence of these records—a 'time-series', as economists would call it—we can examine it for recurring cycles. One cycle stands out straight away: in nearly all parts of the globe, each high tide is followed by another every twelve and a half hours —more or less. But the recurrence is not perfect: the tides do not succeed one another with clockwork precision, and any one high will occur slightly before or after the average—as worked out on the twelve and a half-hour cycle alone. Still, we can now look at these deviations from the average, and see whether there is any cycle to be found there. Do particular high tides happen before or after the average time in an unpredictable way, or is there another regularity here, too? Finding an average cycle in these deviations, we can next examine the departures from this fresh average. And so on.

It is the same with the heights of the tides. Again, there is a basic regularity: the heights of the tides increase towards 'springs' and then decrease towards 'neaps', the whole cycle being completed on an average every fourteen and three-quarter days. But the top springs and bottom neaps are themselves of variable heights: one high spring tide will be followed by another not-so-high spring, which will be followed by a high one, and so on. And the highest springs will commonly occur each year near the time of the equinoxes—i.e. in late March and late September. So, on top of the basic fourteen and three-quarter day cycle, there are further height-variations, on a smaller scale, recurring every twenty-nine and a half and ninety-one and a quarter days.

In either case—whether we study the heights or the times of high tide—we proceed in the same way. First we analyse the records to find the basic cyclical changes; then we study the departures from the average, to find what cycles there are in these deviations; then we study the deviations from the average deviations; and so on. In this way, we progressively bring to light all the cyclical changes which combine to determine the heights and times of the tides.

Now comes the key move. Once we have discovered these cycles from a study of the past records, we can begin to look ahead: we can 'extrapolate'—extend each of the cycles independently ahead to some future date—calculate the respective contributions they will make to the time or height of that day's tides, and combine the resulting figures. Suppose the twelve and a half-hour average indicates a high on a chosen day at 4.37 p.m.; but the cyclical deviation indicates that the high will be seven minutes early that afternoon; while the cycle of deviations from *this* average deviation points to a delay of two minutes. . . . The time to predict will then be 4.37 − 7 + 2, i.e. 4.32 p.m.

This in essence is the technique used to this day for computing the times and heights of tides. Methods of this sort involve, of course, certain assumptions: for instance, the assumption that by carrying the numerical analysis far enough one will be able to bring to light all the different variations involved. The twelve and a half hour cycle is obvious to the simplest observer; the fourteen and three-quarter day cycle becomes clear to anyone who lives for any length of time by the ocean; the smaller-scale and longer-term factors reveal themselves only as a result of longer experience or more careful study. By taking the process far enough, one might hope, it should be possible to compute the times and heights of actual future tides with any accuracy one pleases. But there is in the case of tides a limit to what we can achieve—a difficulty which the Babylonian astronomers were spared. Some of the factors affecting the tides do not vary cyclically, but irregularly. For instance, strong gales blowing into a narrowing channel can cause a tide to rise far higher than it would normally have done, and by preventing the water from running off delay the tide from turning. (The North Sea floods of January 1953 were an extreme example of this phenomenon.) With this limitation, the technique of analysing tidal changes into independent cycles meets all our practical needs.

One point should be noticed, which is immediately relevant to our understanding of Babylonian astronomy. For the purposes of this sort of numerical analysis, we do not have to *explain* why these cyclical changes occur, or what agencies are responsible for them. As a matter of fact, we do by now understand pretty well how the different tidal cycles are brought about. The twelve and a half-hour cycle is related to the rotation of the Earth, acting in combination with the motion of the Moon: the fourteen and three-quarter day cycle follows the Moon's phases, the highest spring tide (according to a well-known rule-of-thumb) being the fifth high tide after the Full Moon. The Sun's gravitational pull has a subsidiary effect on the tides, also: so the ninety-one-day-or-so cycle can be related to the length of the solar year, four such cycles making up the 365 days. All these cyclical variations make sense in terms of the theory of gravitation. But, for the preparation of tide-tables, this theory is beside the point. It may be intellectually satisfying that we can explain the cycles, but for practical purposes we have to know only that they *exist*.

The Babylonians attacked the problems of celestial forecasting using similar techniques. They had lengthy records of eclipses, and as time went on kept progressively more complete records of other astronomical events also. Looking back, they were able to spot more and more of the cyclical variations which combine to determine where precisely in the sky the Moon, the Sun, or a particular planet will appear on a given day or night. They applied the resulting numerical analysis—by extrapolation—to predict eclipses, first visibilities, and stationary points in the planetary tracks. In astronomy, no disturbing factor frustrated their work (as the wind can do with tidal forecasts) and over the centuries we find their skill at celestial forecasting becoming more exact. Locusts and earthquakes, on the other hand, behaved less predictably, and in those directions the same methods of forecasting never had any success.

There is no room to analyse here in full all the cyclical variations in the motions of the Sun and Moon, nor the precise methods the Babylonians used for prediction. But we can indicate some of the difficulties they had to overcome. Consider, for instance, what has to happen for the Moon to be eclipsed. To begin with, the Sun and Moon have to be in precise 'opposition'—that is, 180° apart measured along the ecliptic. This, of course, happens in the middle of every lunar month, at Full Moon. However, the speeds at which the Sun and Moon move across the sky are not quite uniform; so that calculating the moment of opposition is itself a complicated task. And even when this is done, one still has to find out how near the Moon will be at that moment to the line of the ecliptic. For an eclipse to occur, the Moon must be precisely *on* the ecliptic at the moment of opposition, and not (say) ½° or more above or below it: in fact, as we know and the Babylonians knew, the Moon does not keep exactly on the Sun's path, but swings alternately above and below it up to 5° either way.

To predict a lunar eclipse, accordingly, one must have sorted out and recorded separately all the foregoing cycles. These vary independently of one another; nevertheless, one can learn to combine them, so as to yield the desired prediction; and the great insight of the Babylonians is shown by the fact that they recognized this. In fact, they worked out ways of forecasting lunar eclipses infallibly. They were even able to estimate roughly how complete any lunar eclipse would be. And eventually they solved also the trickiest problem of all: that of determining exactly, beforehand, on which evening the crescent of the New Moon would be visible for the first time—so that the legal and religious month could begin.

The problem of predicting *solar* eclipses with certainty defeated them: as we know now, these eclipses are visible only from a narrow band of the Earth's surface, owing to the small shadow the Moon casts on the Earth. Still, they were able to avoid ever being taken by surprise by a solar eclipse. They knew that these always occur exactly at the moment of 'conjunction', when the Moon is completely invisible, being dead in line with the Sun. By studying the path of the Moon as it crosses alternately to one side or the other of the Sun's own track, they could at any rate see when it was possible—or out of the question—for there to be an eclipse of the Sun. (If there was the 'danger' of an eclipse and this did not occur, that itself was regarded as a good omen.)

In talking about the shadow the Moon casts on the Earth, we are explaining *in our own terms* why it was that Babylonian techniques worked in some cases and failed in others. But in doing so we probably go outside their framework of thought. Did they recognize the *cause* of eclipses? We have nothing to show for certain whether they did or not. What they achieved in their own line was striking enough, and it is important to see just how far one can go simply by analysing phenomena—whether tidal or celestial—without bringing in those theoretical questions which are *for us* an essential part of oceanography and astronomy. All that is needed for astronomical predictions such as theirs is to consider what we *see* in the sky. Even the question whether the Earth is moving or not is irrelevant: still more so, questions about 'shadows', 'forces of attraction', and the rest.

The precision achieved by the late Babylonian astronomers, for instance Kidinnu in the late fourth century B.C., has been summed up as follows:

> Hansen, most famous of lunar astronomers, in 1857 gave the value for the annual motion of sun and moon 0·3″ in excess; Kidinnu's error was three times greater. Oppolzer in 1887 constructed the canon we regularly employ to date ancient eclipses. It is now recognized that his value for the motion of the sun from the node was 0·7″ too small per annum; Kidinnu was actually nearer the truth with an error of 0·5″ too great. That such

accuracy could be attained without telescopes, clocks, or the innumerable mechanical appliances which crowd our observatories, and without our higher mathematics, seems incredible until we recall that Kidinnu had at his disposal a longer series of carefully-observed eclipses and other astronomical phenomena than are available to his present-day successors.

The Greeks, whose astronomy is full of speculations about the physical character of the stars and planets and the causes of their motions, did not achieve comparable standards of accuracy until their very last phase. By this time, men such as Hipparchos and Ptolemy could take advantage of the Babylonian work, adapting its results to their own more geometrical methods of thought. For, after the capture of Babylon by Alexander the Great, the Mesopotamian and Greek astronomical traditions were free to flow together—as they eventually did, in Alexandria, from about 200 B.C. on.

The allusion in this last quotation to 'a long series of carefully-observed phenomena' calls for comment. For the precision the Babylonians achieved sprang, not from any great accuracy in their individual observations, but from the antiquity and continuity of their records alone. The few instruments they used were not accurate: simple direction-finders based on the same principle as the sundial. Further, many of the happenings they were most interested in took place on or near the horizon, which must often have been hazy or obscured by sand. Yet given a sufficiently long sequence of records as a basis for extrapolation, one need not be seriously worried either by random fluctuations (such as the wind produces in the tides), or by inaccuracies in the individual observations. The best test of a method of extrapolation is to ask about the proportionate errors it involves: 'How many minutes *per year* will our predictions be out?' By this standard, errors of minutes or even hours in one's records may not be significant when the events in question took place centuries earlier. For an error of twenty-four hours taken over a period of 250 years (i.e. 5.76 minutes per year) is *proportionately* smaller than an error of one hour taken over a period of ten years (six minutes a year). The Babylonians had the two most important things—continuity in their records, and a reliable calendar by which to interpret them.

Looking back at Babylonian astronomy from the twentieth century, one is struck by two things: the care with which the records were kept, and the mathematical brilliance of the predictive techniques. Eventually, science was to owe a great debt to the Babylonian astronomers, for speculative theories about the Heavens could, in the long run, be tested only by seeing how far they explained the observed motions of the heavenly bodies. The Babylonian material was to be fundamental for Hipparchos and Ptolemy. One might almost think—reading history backwards—that the Babylonians had compiled their astronomical records and worked out their analyses of the phenomena *with the intention of* providing the basis for a scientific astronomy.

Yet this seems to be far from the truth. Although they were able to make forecasts of great accuracy, they did so in a way which did nothing to explain the events in question. Their work made eclipses, conjunctions, and retrogradations predictable, but it made them no more *intelligible* than before. One could go on thinking of the planets in any way one pleased, and the regularities in their motions remained fundamentally mysterious. The demand for explanations of natural happenings originated, indeed, not in Babylonia so much as in Greece; and when we ask what a scientific explanation can, and should, do for us, we must bear this multiple origin of science in mind. Tables of planetary positions are computed to this day in the Nautical Almanac Office by empirical methods, just like tide-tables, or the Babylonians' own ephemerides; and these methods even now owe little to the physics of Newton. Our theories may help us to understand why such techniques are effective, but the actual procedures of computation are justified because they work, not because they are theoretically respectable.

Why, then, did the Babylonians *not* theorize about the Heavens? Here the wider connections of their work are relevant. All the chief heavenly bodies were regarded as Gods: even on tablets corresponding closely to our own *Astronomical Ephemeris*—columns of figures recording the positions of the planets day by day—we still find them named by their regular Divine names. The same star-gods are referred to in the traditional creation-myth, which was handed down with only slight alterations for many centuries, and was part of the Holy Scripture of Babylon.

This is the great cosmological poem, or hymn, known as *Enuma Elish*, which has been largely reconstructed from independent fragments discovered by archaeologists during the last century. The first known versions of the myth date from as early as the time of Hammurabi (1800 B.C.); but it never changed very much in form, especially after 700 B.C. It is in part a Creation story, comparable to the *Book of Genesis* in our own Scriptures. But in its canonical form it also has the task of explaining, and so justifying, the positions of Marduk as the chief God of Mesopotamia—he being its nominal hero—and of Babylon as the chief city of the region. Throughout the myth, the two Orders of things, in the Heavens and on the Earth, are presented as being inseparably connected. Babylon (so the myth tells us) was built to Marduk's order by the Gods of the stars, under the command of Anu, the God of the Upper Heavens and the ecliptic.

The opening of the poem is in the literary style familiar to us from the *Psalms*: the lines come in pairs, which repeat one another in slightly different words. In the beginning was Chaos: an undifferentiated expanse of water, arched over by a featureless sky.

When the upper heavens had as yet no name,
And the lower heavens had not as yet been named,
When only the primeval Apsu which was to beget them yet existed,
And their mother Ti'amat, who gave birth to them all;
When all was as yet still mixed in the waters,
And no dry land could be seen—not even a marsh;
When none of the Gods had yet been brought into existence,
Or been given names, or had their destinies fixed:
Then were the Gods created between the Begetters.

Later in the poem comes the stage at which the Order of the Heavens is established. The movements of the Gods are to serve as a natural measure of Time, and the Moon's phases are started.

Then Marduk created places for the Great Gods.
He set up their likenesses in the constellations.
He fixed the year and defined its divisions;
Setting up three constellations for each of the twelve months.
When he had defined the days of the year by the constellations,
He set up the station of Nibiru [the Zodiacal band] as a measure of them all,
That none might be too long or too short,
And set up also the stations of Enlil and Ea [the higher and lower heavens].
He created open gates on both sides,
With strong locks to the East and to the West.
And in the centre he fixed the zenith.
He caused the Moon to shine forth; and put the night under her command.
He appointed her to dwell in the night and mark out the time;
Month after month unceasingly he caused her disc to grow.
'At the beginning of the month, as thou risest over the land,
Thou shalt shine as a horned crescent for six days;
And with half a disc on the seventh day.
At the full moon thou shalt stand in opposition to the Sun, in the middle of each month.
When the Sun has overtaken thee on the eastern horizon,
Thou shalt shrink and shape thy crescent backwards.
As invisibility approaches, draw near to the path of the Sun.
And on the twenty-ninth day thou shalt stand in line with the Sun a second time.'

Finally Marduk orders Ea to create Man, whose task will be to serve the Gods 'in order that they may take their rest'. In return, the Gods build the city of Babylon, including Esagila, the great *ziggurat* (or stepped temple) which was the home of Marduk and the centre of the Babylonian religion.

After Ea, the wise, had created mankind
And had imposed on him the duty of serving the gods.
(That creation was beyond human understanding,
And was performed in accordance with the ingenious plans of Marduk.)
Marduk, king of the gods, divided
The whole population of the Anunnaki [the Gods] both above and below;
And he subordinated them to Anu, so that they should observe his decrees.
Three hundred he set as guards in the Heavens.

And he defined the tracks of the Gods of Earth also.
In Heaven and Earth he caused six hundred to dwell.
After he had issued all his decrees,
Allotting to the Anunnaki of Heaven and Earth all their portions,
The Anunnaki opened their mouths
And cried to Marduk, their lord:
'Now, O Lord, who hast brought about our deliverance from toil,
What shall we do to show you our gratitude?
We will build a shrine whose name shall be called
The Resting-Place for Night—come, let us rest in it.
We will create a shrine, ...;
On the day of our arrival [New Year's day] we will rest in it.'
When Marduk heard this,
His face shone brightly as the day, and he said:
'So let Babylon be, which Ye have desired to construct;
Let a city be built and a well-girt shrine be erected.'
Then the Anunnaki worked with their spades
And shaped bricks for a whole year;
And when the second year came,
They raised the top of Esagila on high, above the Apsu.

The close association of practical astronomy with traditional mythology had a conservative effect in two ways. On the one hand, it made observations of the heavenly bodies a pious duty, and provided an additional motive for collecting and preserving astronomical records; yet at the same time it surrounded the sky and the stars with an aura of sanctity which removed them from the scope of rationalistic speculation. Fate was decreed by the Gods. But the paths of the Gods formed a pattern in the sky which the intellect could master; and to read omens—whether astronomical ones or others—was one way of figuring out the pattern of Fate. Against this background, the arts of divination and astrology were not a quaint kind of delusion, but rather a genuine challenge to human ingenuity. For the Babylonians, the Sun, Moon, and stars determined not only how the tides rose and fell, and how the seasons changed, but equally how their personal and social fortunes would turn out. To reject this belief demanded a comprehensive understanding of the interconnections in Nature such as was achieved only much later. To begin with, there was no essential difference between astrological prediction and astronomical prophecy; and as late as the seventeenth century A.D. professional astronomers like Kepler were prepared to act as astrologers also.

Yet, though Babylonian astronomy was connected with religion and mythology, one can put too much weight on this fact. To say (for instance) that the astronomical records we have recovered were kept in 'temples', and compiled by 'sacred colleges of priests', may be misleading. For this assumes a separation between sacred and secular affairs that is inappropriate when applied to the early cities of Mesopotamia. Religious and political affairs were then interlocked in a way which would be inconceivable now. The art of cuneiform writing was possessed by only a few, and on them devolved tasks which nowadays are divided between the Church, the colleges, and the Civil Service. A reliable calendar was needed, not only for purposes of ritual, but also for civil administration; and from what little evidence we have, we can say only that the astronomers of Babylon were probably state employees. How they were trained, where they worked, whether they had any other functions: none of this we know. Possibly they were attached to the great *ziggurat* which was the religious and administrative centre of the city. (Plate 2.) We may call this building a 'temple' if we like, for it was that among other things. But it would be a mistake to imply that the prime motive of Babylonian astronomy was necessarily religious, when so many other practical motives—and even scholarly ones—existed alongside the religious ones.

In any case, to be honest, we do not know for certain whether the astronomers had any theoretical ideas about the heavenly bodies or not. All we can say is that their practical achievements did not *require* any theoretical insight, and that scarcely a trace has survived of any theories they may have had.

One of the rare pieces of evidence, and a very indirect one at that, is contained in the following passage from Vitruvius. He reports a tradition that Berossos, a priest of Bel (i.e. Marduk), migrated to the island of Cos (c. 290–270 B.C.) and lectured on Babylonian science. Amongst other things he taught a doctrine about the cause of the phases of the Moon. At first sight this may seem to discredit our generalization about Babylonian astronomy—namely, that it was purely computational, and that theoretical speculation played a negligible part. On closer examination, however, the passage rather seems to bear out rather than to falsify this claim; for the theory put forward is, by Greek standards, amazingly primitive:

> Berossos, who came from the city or nation of the Chaldeans and expounded Babylonian science in Asia Minor, taught as follows:
>
> The globe of the Moon is luminous on one hemisphere, the other half being dark blue in colour. When in the course of its journey it comes below the disc of the Sun, the rays of the Sun and its violent heat take hold of it and on account of the properties of light turn the shining half towards that light. But while those upper parts look towards the Sun, the lower part of the Moon, which is not luminous, is indistinguishable from the surrounding atmosphere and so appears dark. When it is quite perpendicular to the rays, all its light is retained on the upper face, and then it is known as the first [or new] moon.
>
> When, moving on, the Moon travels towards the eastern parts of the sky, the action of the Sun on it is weakened, and the very edge of its luminous hemisphere casts its splendour on the Earth in the form of a very thin arc; from which it is called the second moon. Moving round day by day, it is called the third moon, fourth moon and so on. On the seventh day, the Sun being in the West, the Moon occupies the middle of the visible sky and, being half-way across the sky from the Sun, turns half of its shining face towards the Earth. But when on the fourteenth day the whole width of the heavens separates the Sun and the Moon, the Moon rising in the East just as the Sun sets in the West, the Moon is at that distance free of the effects of the Sun's rays and shows the full glory of its whole sphere as a complete disc. During the remaining days until the completion of the lunar month it diminishes daily, turning as it comes once more under the influence of the Sun's rays, and so effecting the days of the month in due order.

This explanation ignores half a dozen theoretical advances made by Greek thinkers in the two centuries preceding 270 B.C. So long as we think of the Sun and Moon as going round closed orbits, we can immediately attribute these phases—as the Greeks did—to the changing angle at which the Moon reflects the Sun's light. But Berossos does not even recognize that the Moon's light is borrowed;

and his whole picture supposes that the two bodies go in straight-line paths at different heights across the sky above the flat Earth. If we retain the traditional picture—reminiscent of *Enuma Elish*, with its

'open gates on both sides
With strong locks to the East and to the West'

we can understand Berossos' theory. But, to a Greek, it must have seemed terribly crude.

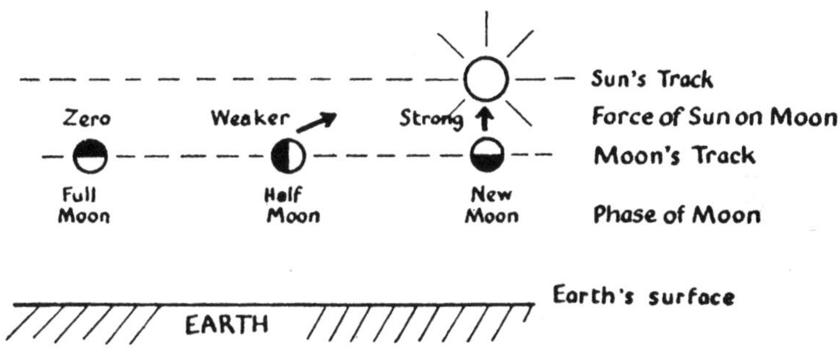

Berossos' theory of the Moon's phases

To sum up. What remains of Babylonian astronomy is the relics of a long tradition of arithmetical brilliance, harnessed to the problem of celestial forecasting. Presumably the scholars and scribes who made a living by preparing tables of ephemerides used to discuss much more than they ever committed to clay. But if they did so, their ideas never became common property, and are still lost. Certainly they did not react back on the general mythological tradition.

This fact is significant in two ways, and shows how far the Babylonians were from doing 'Science', in the full twentieth-century sense of the word. As we see it, the life of science depends on people's having the chance, not only to think for themselves, but also to pass their insights on to others. It is not enough just to hug a great new idea to oneself, or to justify it to a few professional colleagues. An intellectual revolution is complete only when the fundamental insight on which it rests has become part of the general heritage of 'common sense'.

More specifically, this insulation of Babylonian mythology from astronomical criticism takes us back to a phase in thought when cosmology and astronomy were still as distinct as they were for the author of *Genesis*. Our contemporary ideas about the structure of the universe spring from a theoretical tradition which leads back, not to Babylon, but to Greece: the early Greek philosophers were the first to realize that cosmological ideas must stand up to criticism when tested against the 'appearances' in the Heavens. Yet their own astronomical records were scanty. So—though the Greek marriage of astronomy with cosmology might have seemed to devout followers of Marduk an illicit union—it was to the Babylonian records that the Greeks had eventually to resort as a touchstone for their own theories. We must turn our attention next to this other, more speculative branch of science's family-tree.

NOTE: HOW THE BABYLONIANS COMPUTED CONJUNCTIONS

As an example of the arithmetical procedures the Babylonians used when making their astronomical computations, consider the following table:

2,59	I	28,37,57,58	20,46,16,14	Taurus
	II	28,19,57,58	19, 6,14,12	Gemini
	III	28,19,21,22	17,25,35,34	Cancer
	IV	28,37,21,22	16, 2,56,56	Leo
	V	28,55,21,22	14,58,18,18	Virgo
	VI	29,13,21,22	14,11,39,40	Libra
	VII	29,31,21,22	13,43, 1, 2	Scorpio
	VIII	29,49,21,22	13,32,22,24	Sagittarius
	IX	29,56,36,38	13,28,59, 2	Capricorn
	X	29,38,36,38	13, 7,35,40	Aquarius
	XI	29,20,36,38	12,28,12,18	Pisces
	XII	29, 2,36,38	11,30,48,56	Aries

This table is a direct transcription into our notation of a cuneiform tablet referring to the year 133–2 B.C. The date is determined by the figures 2,59 at the left-hand side: these denote the year 2,59 of the Seleucid era, i.e. $2 \times 60 + 59$ (=179) years after 312 B.C., the beginning of the era. (In sexagesimal notation, 2,59 corresponds to our own decimal numeral 179.)

How is the table to be interpreted? It apparently gives computed positions in the sky at which 'conjunctions' between the Sun and the Moon were to be expected during the year in question. (See page 39 above.) The column of figures I to XII denotes the successive months of the year. The figures in the next column, beginning 28,37,57,58 (i.e. 28° 37′ 57″ 58‴), give the calculated amounts which the Sun would travel across the Zodiac each month, between one conjunction and the next. Notice how the figures in the last pair of places in these sexagesimal numerals repeat themselves: they all end either 57,58 or 21,22 or 36,38. It is inconceivable that actual observations of the movements of the Sun should show such neatness; so these repetitions confirm that we are here dealing with computed, not observed, movements. The final column of figures gives the anticipated positions in the Zodiac of the conjunctions in each of the twelve months. Each position listed in the right-hand column is obtained by adding to the position for the month before (the figure immediately above it) the amount of the 'monthly progress' given in the left-hand column: every time a new sign of the Zodiac is entered, one subtracts 30°. For instance, add the figures in the third line down of the left-hand column to the figures in the second line line down of the right-hand column, and the sum is the figures in the third line down of the right-hand column: the month's motion of 28° 19′ 21″ 22‴ takes the Sun on from the position 19° 6′ 14″ 12‴ in the sign of Gemini to the position 17° 25′ 35″ 34‴ in the next sign, Cancer. (Note: 28° + 19° =47° in Gemini corresponds, after subtracting 30°, to 17° in Cancer.)

On what rule is this calculation based? The fundamental assumption is, that in any month the Sun moves on by an amount 18′ (0·3°) either more or less than it did in the previous month. Between months V and VI, for instance, it is listed as moving 29° 13′ 21″ 22‴: between months VI and VII it is allowed 18′ more, i.e. 29° 31′ 21″ 22‴. It accelerates month by month until it comes close to a maximum rate of 30° 1′ 59″ 0‴ per month, then slows down again until it approaches a minimum rate of 28° 10′ 39″ 40‴

per month. We would nowadays represent the Sun's speed across the Zodiac by a wave-like graph: the Babylonians approximated to this wave by using a pair of straight lines—one down, one up. (See diagram) You can check for yourselves how they made the

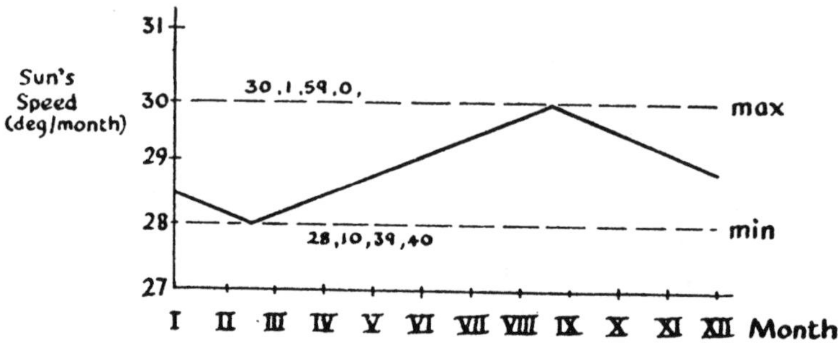

Sun's speed 'turn the corner', whenever adding or subtracting a full 18′ would cause the rate either to exceed the maximum estimated rate, or to drop below the minimum. As a hint: the changeover from figures ending 57,58 to figures ending 21,22 represents a passage past a minimum; and the subsequent changeover to figures ending 36,38 represents the next passage past a maximum.

For a full explanation of this table, consult O. Neugebauer, *The Exact Sciences in Antiquity*, pp. 101–10.

FURTHER READING AND REFERENCES

The most valuable single book for anyone interested in Babylonian astronomy and mathematics is

O. Neugebauer: *The Exact Sciences in Antiquity*

Early Mesopotamian science is discussed in its historical background in

A. T. Olmstead: *History of the Persian Empire*

For the general historical background of science in the Ancient Empires, the following are useful:

V. Gordon Childe: *What Happened in History*
J. H. Breasted: *Ancient Times*
Leonard Woolley: *The Sumerians*

The mythological aspects of pre-scientific thought are set out and interestingly discussed in

H. Frankfort and others: *Before Philosophy* (first published as *The Intellectual Adventure of Ancient Man*)

The books by Childe and Frankfort are included in the Pelican paper-back series, and Olmstead's in the Phoenix series.

The 16-mm. film *Rivers of Time*, made under the auspices of the Iraq Petroleum Company, gives a splendid visual introduction to the Sumerian and Arab civilizations in Mesopotamia.

The first part of the film *Earth and Sky*, issued in conjunction with this book, deals with Babylonian astronomy: retrograde motion and cuneiform writing are both demonstrated in this section.

2

The Invention of Theory

TURNING from Babylonian astronomy to early Greek speculations gives the impression of passing from one intellectual discipline to another completely different one. And this is in effect what has happened. The first Greek attempt at astrophysical theory no more amounted to 'science', as we know it today, than did the Babylonian techniques. Yet each of them independently contained a germ which, later, was to contribute to the development of the scientific tradition. Both of them belong in the ancestry of science.

The whole aim of Babylonian astronomy was to serve as an instrument of prediction and divination—to forecast the astronomical appearances, rather than to make sense of them. The fundamental problem motivating the Greek natural philosophers was quite different. Whether or not one could predict all the changes in the Heavens or on the Earth, at any rate one should try to make out, in general terms, why these changes happened as they did. Night and day, summer and winter, youth and age, sickness and health: all these succeeded one another like the images on a cinema screen. But behind the 'flux' of everyday happenings (the Greeks were convinced) lay eternal, permanent 'principles'. If only one could discover 'the nature of things' one would be able to explain, in a rational manner, much that would otherwise merely perplex.

THE SOURCES

Our knowledge of early Greek science or 'natural philosophy' (the terms are interchangeable) is derived from sources which have never been completely lost to human knowledge, as the Babylonian cuneiform tablets were lost. Some of the most important works, it is true, were unknown in Western Europe for something like a thousand years: copies of the original texts—or copies of copies—were preserved only in the libraries of Constantinople, or else in Arabic translation in Cordova, Cairo, and Baghdad. But some knowledge of Greek ideas was always preserved, and scholars have progressively increased the range of authors whose work we can study.

The centuries have dealt more kindly with some authors than with others. With very few exceptions, the works of Plato and Aristotle have come down to us complete. In their case, the problems are, rather, to sift out the genuine works from the spurious ones that circulated during the Middle Ages under their august names; and to decide, in the light of various manuscripts still preserved, how the works originally read. The final encyclopaedia of Greek astronomy, compiled by Claudius Ptolemy around A.D. 150, has also come down to us complete, though Hipparchos (to whom Ptolemy owed a great deal) is known only from fragments and second-hand reports.

Many other authors whose writings would have interested us here have suffered the same fate as Hipparchos. Before the time of the great Athenian philosophers, original speculation about Nature was centred in two regions, both of them outside the boundaries of present-day Greece: Ionia, the coastal strip of Asia Minor facing the Aegean Sea (particularly the town of Miletus), and the Greek towns of Sicily and Southern Italy. We can reconstruct these 'pre-Socratic' traditions only from fragments and hearsay, and these are often reported by hostile critics. Enough remains to show us the general sorts of theories these men put forward, but no more: if we press for exact interpretations, the answers the fragments give us are merely tantalizing and ambiguous. This is a great pity. For in the development of scientific ideas, the ideas of the minor figures can be as revealing as those of the giants; and we would understand better the character of the Greek tradition in the scientific field, if more of their work had survived. Still, enough remains even so to provide the outlines of an intelligible picture.

Before leaving the sources, one thing should be added. We in the twentieth century have at our disposal a good deal more material from the Greek philosophers than our predecessors in (say) the seventeenth century, and far more than the scholars of the fourteenth century, to say nothing of the eleventh century. Further, the editions we can now work with are superior; and the whole picture we can reconstruct of Greek thought is more detailed, more exact, and in better proportion. This has especially to be borne in mind when one considers (for instance) the justice of seventeenth-century attacks on the views of Aristotle or Ptolemy or Galen. For, as a result of centuries of handling by scribes, translators, and commentators, the theories of these men had been partly elaborated and developed, and partly corrupted: one has to distinguish carefully between the Aristotelian or Ptolemaic or Galenic traditions, against which these tirades were immediately directed, and the original views of the authors themselves.

Here, of course, we are concerned with the problems actually facing the Greeks when they first tried to make sense of the workings of Nature, and the novel theories they put forward to account for them. The changes which their ideas underwent in the subsequent 2000 years will be a later part of our story.

THE BACKGROUND

In the centuries before 300 B.C. the Greek lands around the Aegean Sea stood—both politically and in many other ways—in striking contrast to Mesopotamia. Though the balance of power between the Rivers Tigris and Euphrates shifted between the Assyrians and the Babylonians, the Medes and the Persians, the region remained throughout this time part of a large, centralized land-empire. Dynasties might rise and fall, the land might be politically united with Persia to the East, or with Syria and Anatolia to the West; but otherwise life went on in much the same way. Social organization in the region had first grown up in the Sumerian period, well before 3000 B.C., around the great network of irrigation-canals which spread from the deltas inland. This network continued in operation equally under the Babylonians and Persians—and later under the

Macedonians, Romans, Parthians, Sassanians, and Arabs; being destroyed only as a result of the Mongol invasion in A.D. 1258. Trade and communications continued with little interruption. Mythologies were passed down from generation to generation. Tribute and taxes had still to be paid. Religious rituals continued on their traditional cycles. The demand for an accurate calendar and for skilful divination remained pressing, so providing work for a class of professional astronomers.

Around the Aegean, on the other hand, there was no such order and stability. Ionia was on the very circumference of the Middle Eastern land-empires, and Miletus remained for centuries a bone of contention between the Greeks and the Persians. Many of the Greek cities were ports, dependent for their prosperity on maritime trade, both with Egypt and the Eastern Mediterranean and with their own outposts or 'colonies' in Italy and further west. (Plate 3.) After the breakdown of the Minoan Empire centred on Crete and of the Mycenaean power on the Greek mainland (whose exploits in the centuries before 1000 B.C. are recorded in the legends of the Trojan War), the cities of the Greek world were effectively independent of one another for many centuries: until 338 B.C. they were never united politically, but at most collaborated together in a loose confederacy. As the fortunes of trade or war tilted one way or another, one city would rise into prominence, another would drop into obscurity. As a result, it was a world, not only free of any centralized authority but also without any long political or intellectual traditions.

Against this background, the main cultural differences between Mesopotamia and the Greek world are readily intelligible. Under the Babylonians and the Persians alike, Mesopotamian society was essentially conservative. On the positive side, its archives preserved down the centuries the records of tradition, not only in the fields of law and religion, but also in mathematics and astronomy; but equally, there was nothing about the situation to stimulate intellectual ferment or to encourage original speculation or heterodox ideas. The situation in the Greek world was rather different. True, the Greeks had their own traditions in poetry and religion, and the Gods of Olympus were not very different in character from contemporary Gods in the Middle East. But, quite early, some of the Greeks could not help being struck by—for instance—the central problems of comparative religion.

They were, of course, especially well placed to appreciate these problems. The trade routes overland from Mesopotamia came down to the coast at the Greek ports; commerce led to personal contact with the Phoenicians, the Egyptians, and the Etruscans; and through these people came also second-hand reports about the beliefs of other peoples still further afield. How, then, were the Greeks to reconcile the Egyptian tales of Thoth and Osiris, or the Mesopotamian myths about Marduk and Ea, with their own traditions about Zeus, Apollo, and the rest? We can well understand the reaction of Xenophanes, who was born at Kolophon in Ionia about 560 B.C. but migrated to Sicily about the age of twenty-five. Faced with the conflicting traditions of different peoples, he dismissed all mythologies equally—the Greek included—on the grounds that they were excessively anthropomorphic:

> Homer and Hesiod have ascribed to the gods all things that are a shame and a disgrace among mortals: theft and adultery and mutual deceit. Mortals consider that the gods are begotten as they are, and have clothes and voices and figures like theirs. The Ethiopians make their gods black and snub-nosed; the Thracians say theirs have blue eyes and red hair. Yes, and if oxen and horses or lions had hands, and could paint with their hands, and produce works of art as men do, horses would paint the gods with shapes like horses, and oxen like oxen, and make their bodies in the image of their several kinds.

It is perhaps understandable that these ideas originated at a meeting-point of different cultures. With great originality, Xenophanes went on to invoke a single, omnipresent God, 'neither in form like unto mortals nor in thought', who 'without toil swayeth all things by the thought of his mind'. And alongside this novel monotheism, he put forward some ideas about astronomy and meteorology. The heavenly bodies (he argued) are not Gods, but rather luminous clouds; and 'she that they call Iris [the rainbow] is a cloud likewise, purple, scarlet and green to behold'. Here, in the mercantile cities of the Greek Aegean, we find critical, dissatisfied curiosity and the demand for rationality beginning to yield an intellectual harvest of a new kind.

Yet the situation had a negative side also. Without any tradition of astronomical record-keeping or calculation, there was no demand for professional astronomers. The natural philosophers of ancient Greece were pure intellectuals: a few of them earned a living as teachers or doctors, but most of them were men of leisure. Some historians have placed great emphasis on the political and economic basis of Greek thought, claiming that mercantile democracies have at all periods given positive encouragement to original speculation. This is going much too far: critical speculation about the powers of Nature was in fact an unpopular, minority activity even in Athens. Socrates was criticized at his trial—and pilloried by Aristophanes in his play *The Clouds*—for his allegedly undue curiosity about the nature of celestial things. (Yet Socrates, as we know, soon abandoned his early interest in scientific questions, and devoted himself to the problems of politics, personal conduct, and immortality.) Anaxagoras, too, was imprisoned for impiety and later expelled from Athens—under the great Pericles himself—for teaching that the Sun was a red-hot rock, and the Moon made of earth.

Still, conditions in the Greek world did provide *some* stimulus to original thinking, and social life was not so rigidly ordered as to rule out completely the public discussion and dissemination of heterodox ideas. But, all in all, in the five centuries following 650 B.C., the number of people who contributed actively to 'Greek science' can have been only a few dozen; and the number of their compatriots who read or listened to their teachings with any real understanding probably amounted to no more than a few hundred. This is something one has to remember when assessing their achievements, and it is a mistake, accordingly, to look for too great a discontinuity between Greek speculations about Nature and what had gone before.

For our purposes, the interesting aspects of their work are the forward-looking ones, the seeds from which key ideas in later science eventually descended. At the same time, the forms of later scientific theory were slow in developing: one begins to feel really at home only after about 300 B.C., with the work of (e.g.) the mathematicians Euclid and Archimedes. Many of the earliest and most influential speculations were presented as 'rational myths'; some of them were actually written as poems; and, in at least one case, a highly significant idea first appeared as the private doctrine of a religious brotherhood. This was the embryonic mathematical physics of the Pythagoreans. Pythagoras, it is clear, was not so much the leader of a scientific research team, or the principal of an educational establishment, as (in modern terms) the *guru* of an Indian *ashram*. His recognition of the mathematical basis of natural phenomena was part of a religious revelation; and the doctrine was taught first only to those who had been initiated into his sect. Brotherhoods like this one, and the few 'academies' which grew up around great individual teachers, were the only places where the early Greek philosophers could avoid complete intellectual isolation.

After 300 B.C., as the intellectual focus began to shift to Alexandria, the political authorities at last began to patronize intellectual activities. But this was the patronage of an enlightened monarchy. Greek democracy might to some extent tolerate natural speculation, but it did little to promote it. Rather, its attitude to the new thinkers appearing in its midst was like that of the authorities in Revolutionary France: *La République n'a pas besoin de savants.* However, Greek science in the Macedonian and Alexandrian periods, from Aristotle on, will concern us in a later chapter. For the moment, we shall be looking at the founders of Greek natural philosophy: the Ionians, the Italians, Plato, and Eudoxos.

THE CHARACTER OF GREEK THEORY

The early Greek philosophers described the underlying nature of things in many and varied terms. Some of them echoed the traditional idea that the primeval state of the universe had been a watery Chaos 'without form and void': they envisaged a basic stuff which, like water, was capable of taking many different forms, and from which all the objects of the world might be constituted. Others rejected the idea of a single basic stuff, in favour of three or four—mixed in different proportions. Some invoked 'atoms' as the basis of all things, though these were very different from the twentieth-century physicist's atoms. Others saw the principles behind phenomena as being mathematical in character, and stated their explanations in purely mathematical terms. Others, again, complained—with some justice—that bare mathematical explanations alone can never answer all our questions about the changing world of colour, sound, and life that we know in our daily lives.

Yet, for all these disagreements, one fundamental conviction united them all. *They were philosophers, not prophets.* Behind the changing flux of experience, unchanging principles existed; and by reflecting about our experience we could bring them to light—whether they turned out in the end to be elementary ingredients, or basic mathematical axioms, or form-giving agencies. The philosopher must persevere in his search for them, though this might even mean denying the evidence of his senses and rejecting tradition. Belief that these principles existed was *a matter of faith*, just as much as the Babylonians' belief in the predictability of stellar phenomena. But this faith began by prompting a few men's curiosity in a radically new direction, so that they started to puzzle for the first time about quite new questions. And from these small beginnings, it grew, established itself, and developed in the Greek world over a period of seven centuries.

By the time of the Roman Empire this faith had transformed and largely displaced the traditional mythology, and it lost its hold seriously only around A.D. 200, with the rise of Christian mysticism and the spread of other Oriental cults from Syria and Egypt. The reasons for this failure we shall have to look at later: they sprang in part from certain fundamental weaknesses in the philosophers' answers to the questions of physical astronomy. Yet in the long run, these deficiencies are unimportant. The Greeks were the effective forerunners of modern science, not in their particular answers and theories, but rather in the new *questions* they put into circulation. They first sketched the programme which modern scientists have taken up and refined. To them belongs the glory of having invented the very idea of a scientific theory.

In one field—geometry—they excelled. Elsewhere, their speculative theories led to curiously little effective result. Again and again, one finds in the Greeks hints of later discoveries without these insights ever being driven home. One is shown that so-and-so *might* be the explanation of a given phenomenon; but never cogently convinced that this *must* be the explanation.

There are two reasons for this. The Greeks were chiefly anxious to show that explanations of the kind they favoured were *possible*, rather than to give this or that exact explanation; moreover, to some of them, being able to put forward a number of alternative theories, any of which would do, actually seemed a good thing. An example of the more extreme position is Epicurus. He was keen to explain all the things that happen to us in natural, rational terms, but his motives were not those of a scientist, so much as those of a 'scientific humanist': 'Don't think that dreams (say) are *mysterious*,' he declares, 'Don't think that you are forced to interpret them as supernatural visitations, or treat them as omens.' (Remember: this was a time when even kings consulted oracles, in order to have their dreams interpreted.) 'Why, there's any number of ways in which dreams could happen perfectly naturally, without our having to attribute them to the Gods. The Gods have better things to do than appearing to men in dreams. So don't worry your head about them, for there is really no need.'

From this point of view—if superstitious fear of the Gods is the enemy—the more alternative natural explanations one can give, the better. If dreams might happen perfectly naturally in many different ways, there was that much less reason to be anxious about them—or so Epicurus could plausibly argue: the larger the number of possible natural theories, the less likely it was that one would be driven back on a supernatural explanation.

Even those philosophers whose motives were scientific rather than religious were scarcely in a position to drive their speculations home to the point of proof. Their first task was to get their ideas stated in a consistent and plausible form; then they had to see how far the results tallied with the observations of common experience; and only if they had got past these two stages successfully could they have proceeded to the stage of strict and critical proof. For the moment, their efforts were concentrated on the first two stages, and material was generally lacking which would have conclusively established the adequacy of one of their explanations and the inadequacy of its rivals.

Should we criticize the Greek philosophers for this? Was it a failure on their part not to have adopted earlier the practice of severe criticism in the light of controlled observation which is the heart of modern science's 'experimental method'? In case one is tempted to be scornful, two things should be said. First: you can start asking which of two automobiles performs better—which has the better acceleration or petrol consumption—only when the rival cars are assembled and in working order: the Greek philosophers, rather, should be compared to those men who first envisaged the possibility of motor-cars, and who worked out for us the original tentative designs. Their theories were still only on the 'drawing-board'.

Secondly: among the Greeks, science was a purely *intellectual* enterprise—one which was not undertaken with any *technological* end in view, and which in this direction yielded only the slightest fruit. Not until the later stages of Greek science did men apply its results to the design of water-clocks, taxi-meters, and the like. (The practical tradition of Greek medicine, too, was effectively independent of the philosophical tradition right up to the time of Galen.) The Babylonian astronomers had their noses kept forcibly to the ground by the practical demand for accurate prophecies and predictions. By contrast, the very fertility and freedom with which the Greeks speculated is connected with the fact that so little hung upon the soundness of its results. Suppose they had had to pay for wild generalization, or unsound theorizing, or incautious analogies a price measured in human lives, or collapsed bridges, or misforecast omens; then they might well have proceeded more cautiously. And if their originality and imagination had been shackled in this way, it would have been very much to our loss.

These, then, are the reasons why the Greek philosophers never quite qualified for our modern title of 'scientists'. It was, of course, essential, then as now, to think up fruitful theoretical ideas. But we have since placed on scientists the additional burden of experimental demonstration—and often trust our lives to the results. It is no longer enough to have the right ideas, as the Greeks in many cases did: one must also elaborate them, test them, and prove them. This effective union of theory and practice, which is characteristic of 'science' nowadays, had to wait for a later age.

What was novel in the Greek way of thinking about the world? Perhaps the best way of putting it is to say: they transformed our whole manner of understanding natural phenomena. By their definition, the philosopher was a man who related the visible changes in Nature to the permanent principles underlying them, showing in this way *why* the events happened as they did. The mythologist or magic-worker might have an intuitive 'feel' for the moods of Nature, a familiarity born of long practical experience, and his advice might often be effective for this very reason—he would be the repository also of folk-remedies, and much other traditional lore. But the natural philosopher was not satisfied until he could back up his explanations by *arguments*. He must have a *theory*.

Now, magic and theory alike take for granted that the world is orderly, but they interpret this orderliness in different ways. Many of us have a false idea about magical beliefs, thinking of the witch-doctor or priest as a man who pretends to work miracles in the name of hidden ghosts or spirits. Actually, the attitude of the witch-doctor is a great deal more concrete than this: he is much nearer in spirit to the craftsman than to the mystic. 'Seeing the world aright' —understanding it—means to him, not having a private telescope into a world above the sky, but rather being on working terms with natural things: knowing them personally, so to speak. His art is to handle our dealings with Nature, rather as a diplomat handles negotiations with foreign powers. He reads the mind of the powers-that-be in Nature, and influences them by ritual or intercession. The philosopher, on the other hand, is convinced that Nature functions, not wilfully, but 'rationally'—according to firm principles. One can theorize systematically about the world; be taught how it works, not just get the feel of it. When we have recognized the principles of things, and how even the Gods are obliged to act as they do, then we can truthfully say we *understand*. When the clouds form and the rains fall, this is not evidence of Divine favour or anger, but the inevitable working-out of natural processes: as Aristotle puts it, 'Zeus [the Sky] does not rain in order to make the crops grow, but from necessity'—i.e. the rain falls when it has to. And the philosopher's essential task in this field is to get an intellectual grasp of the character of natural order.

In the earlier mythologies, cosmology—the description of Nature—took the form of dramatic historical stories; like the Babylonian story of Marduk and the Anunnaki. As has been realized more and more in recent years, the Greeks did not at once abandon this element. Many beliefs originating in the mythological period were retained throughout the Golden Age in Athens (500–300 B.C.). The universe was still regarded by most people as a society in which stars, men, and cities shared in a common order. The gods were still felt to be active in the world around, and the heavenly bodies were numbered among them. Even Plato's chief scientific dialogue, the *Timaeus*, begins with a historical-looking myth, in which the 'Demiurge' performs many of the acts of creation attributed to Marduk in the Babylonian story. But there is one crucial difference. Marduk did what he pleased, but the Demiurge was obliged to order the world on rational principles. As Plato remarks, the myth alone is not enough: there is beside the creation-myth another story about Nature to be told, and this is a theoretical one—which explains the structure and behaviour of natural things in terms of underlying principles.

A modern scientist, looking back at the Greek philosophers, often feels exasperated at the amount of sheer argument and 'logic-chopping', and is tempted to criticize them for not having 'got on with the real job'. To him, the title of Aristotle's *Physics* appears quite inappropriate: scarcely any observations or experiments are reported in the treatise, and it is devoted almost entirely to armchair argumentation. How can such a work—he will ask—have made any serious contribution to the progress of science?

Yet it would be a grave mistake to dismiss the Greek tradition of rational argument about physics as irrelevant to science: all this argumentation was very much to the point—to the point, that is to say, which in their time rightly concerned the Greek philosophers. For, in conducting any examination—whether of students or of theories—one has always, as a preliminary, to scrutinize the list of candidates and see which of them are qualified to take part. Something similar happens in the history of every science, when men begin to put forward theories of a novel sort as being 'the explanation' of this, that, or the other. (Think how many words have been written since 1900 about the legitimacy of psycho-analysis.) So, when men began to consider seriously the possibility of explaining natural happenings in terms of atoms, or numbers, or geometrical shapes, they had first to satisfy themselves that this was a reasonable programme—that, at any rate, there were no unanswerable objections. Faced with so great a variety of ideas, the Greeks had (as we shall see) to rule out for a start those which were internally inconsistent—or led into demonstrable contradictions—or clearly failed to cover even the most familiar everyday experiences. That much was sheer good sense.

We may feel in retrospect that they were sometimes oversensitive to logical objections—that they should have had the courage to go ahead: producing more detailed theories which could readily be compared with the facts, instead of getting bogged down in preliminary arguments about principles. But this is to be wise after the event. It is something to have invented the whole notion of 'theory' at all, and the Greeks were rightly proud of having been the originators of philosophical inquiry. And in any case, it was largely out of this debate about rival *forms* of theory that the chief concepts of science were gradually defined. As a result, ideas which had at first been vague and general at last came to be stated with the precision which is rightly demanded of any satisfactory scientific theory. Thus the central problems of science gradually came into focus.

THE FIRST THEORIES

The first Greek attempts to explain the structure and behaviour of things in the Heavens date from the period 600–450 B.C., and rely largely on arguments from analogy. In comparing the remote objects in the sky with familiar bodies around us on the earth, the Ionians were using the *method of theoretical models*, and they were the first people we know of to make extensive use of this method of explanation. In modern physics many of the properties of light can be explained by comparing them with corresponding properties of water-waves, which are familiar to us and can be seen directly: the Ionians looked for similar comparisons between the Heavens and the Earth. A convincing terrestrial model (they thought) entitles one to suppose an analogous set-up in the sky.

This sort of theoretical thinking can be very fruitful. In the field of astronomy, it led them quickly—though not all at once—to recognize that the Moon's light is reflected from the Sun; and that it is eclipsed when the shadow of the Earth cuts off the Sun's light from it. (As we saw, the Babylonian Berossos did not understand these phenomena even as late as 290 B.C.) Many of their speculations were plausible, and some we still accept today, but others were less happy: until men had some idea of the geometric layout of the stars and planets, there could be no great progress in astronomy. At this stage no one had yet demonstrated the vast depths of the sky, or estimated with any accuracy how far the Sun and Moon are from the Earth. It was, therefore, entirely natural for (e.g.) Xenophanes to compare such things as the Sun and stars (astronomical objects, at a distance) with things like rainbows and shining clouds (atmospheric objects, near at hand). For the same reason Anaxagoras, who recognized the causes of eclipses, also suggested that, at the time of the summer solstice, the Sun is turned back at the Tropic of Cancer—to the south of Egypt—by the colder air of the northern regions.

Traditionally, the first of the Greek natural philosophers was Thales, who was teaching shortly after 600 B.C. But the first man whose doctrines tradition has preserved at any length is Anaximander, who was about a generation younger than Thales. Most of what we know about his ideas have come to us by way of Aristotle's pupil Theophrastos, though Aristotle refers to him in passing.

In the selection of passages reproduced here, the key-word (*Apeiron*) is left untranslated, for it is ambiguous. It combines the meaning of 'unbounded' with that of 'unrestricted' and 'unlimited'. The *Apeiron* is not only devoid of boundaries; it is also capable of taking all possible forms and properties. The usual translation ('the

Infinite') is accordingly quite inadequate. The following passages have been selected in order to indicate his views on the origin of the universe, the nature of the heavenly bodies, and the causes of meteorological happenings.

Anaximander of Miletos, son of Praxiades, a fellow-citizen and associate of Thales, said that the material cause and first element of things was the *Apeiron,* he being the first to introduce this name for the material basis of things. It is (he said) identical neither with water nor with any other of the so-called elementary substances, but is something different from them, which is unconfined and from which arise all the heavens and the worlds within them. . . . In addition, there was an eternal motion, by which was brought about the origin of the worlds.

Anaximander did not explain the origin of things in terms of any material changes, but put it down to opposing properties in the 'unrestricted' substratum separating out. At the beginning of the present world, something was separated off capable of begetting hot and cold out of the eternal. From this arose a sphere of flame which fitted as closely round the earth's atmosphere as the bark round a tree. When this had been stripped off and enclosed in rings, the Sun, Moon, and stars came into existence. As for the sea: at first, the whole terrestrial region was moist; and, as it was dried up by the sun, the portion of it that evaporated produced the winds and the turnings-back of the sun and moon, while the portion left behind was the sea. So—on this view—at last it will all be dry.

Anaximander said that the Earth is cylindrical in shape, its depth being one-third of its breadth. It swings freely, held in its place by nothing, staying where it is because of its equal distance from everything else. It is hollow and round, like a stone pillar: we live on one of the flat surfaces, the other being the opposite face of the cylinder.

In the heavens there are wheels of fire, separated from the fire of the world and surrounded by air. In these wheels are breathing-holes: pipe-like passages, through which the fire is visible [as the Sun, Moon and stars]. When the breathing-holes are stopped up, eclipses take place. And the moon seems to wax and wane as the passages open and close. The wheel of the Sun is 27 times the size of the Earth, while that of the moon is 18 times as large. The Sun's wheel is the highest [i.e. the furthest away] while the lowest are the wheels of the stars.

Thunder and lightning are caused by the explosive blast of the wind. When it is shut up in a thick cloud and bursts forth violently, the tearing of the cloud makes the noise of the thunder, while the rift appears as a lightning flash by contrast with the blackness of the cloud. Wind is a current of air, arising when the finest and moistest particles are stirred or melted by the Sun.

The last of the original group of natural philosophers who taught at Miletos in the sixth century B.C. was Anaximenes. When we turn to look at the development of ideas about the material constitution of things, he will be important as the man who first described the universal underlying substance of things as *pneuma*, or breath: though we may sometimes translate this as Air, Anaximenes clearly does not think of it (as we would do) as a mixture of inert gases, but rather as 'the breath of life' which animates the whole universe. His views, too, have come to us largely through Theophrastos. The following brief selection reports his astronomical and meteorological doctrines.

Anaximenes of Miletos, son of Eurystratos, who had been an associate of Anaximander, said, like him, that the underlying substance was single and unbounded. However, according to him it was not characterless, as it was for Anaximander, but had a definite character; for he said that it was Air.

He said that, while the Air was being thickened like felt, the Earth first came into being. The Earth is like a table in shape. It is very broad, and can accordingly be supported by the Air. The Sun, Moon and other heavenly bodies, which are of a fiery nature, are likewise supported by the Air because of their breadth. The heavenly bodies were produced out of moisture rising from the Earth. When this rarefied, fire was produced: the stars are composed of the fire thus raised aloft. Revolving along with the stars there are also earthy bodies.

The heavenly bodies do not move under the Earth, as some suppose, but round it; like a cap turned round on one's head. The Sun disappears from sight, not because it goes below the Earth, but because, having gone a long way from us, it is concealed by the higher parts of the Earth. The stars give no heat because they are so far away. They are fixed like nails in the crystalline vault of the heavens, though some say they are like fiery leaves painted on the heavens.

Anaximenes explained lightning like Anaximander, citing as an illustration the way in which the sea flashes when it is split by the oars of a boat. Hail is produced when water freezes in falling; snow, when some air is imprisoned in the water. The rainbow is produced when the beams of the

Sun fall on thick, condensed Air. The front part seems red, being burnt by the sun's rays, while the other part is darker, owing to the predominance of moisture. A rainbow can be produced at night by the Moon, but not often, because the Moon is not always full, and its light is weaker than the Sun's.

The most influential of the Ionian natural philosophers was Anaxagoras, who came from Klazomenai, but is said to have taught in Athens for thirty years: he was born in or about 500 B.C. and (as we saw) was banished from Athens around 450 B.C., after which he returned to Asia Minor and taught at Lampsakos until his death about 428 B.C.

Anaxagoras followed, and improved on, the astronomical and meteorological ideas of Anaximander and Anaximenes, being the first of the Ionians to state clearly the causes of eclipses. His views on these subjects can be seen from the following passages:

It is the Sun that puts the brightness into the Moon. We call 'rainbow' the reflexion of the Sun in the clouds. This is a sign of storm: the water that flows round the cloud causes wind and pours down as rain.

The Earth is flat, and remains suspended because of its size and because there is no vacuum. The Air is very strong, and supports the Earth. As for the moisture on the Earth's surface, the sea originated partly from the waters on top of the Earth—for as these evaporated, what remained turned salt—and partly from the rivers which flow into it. Rivers owe their existence both to the rains, and to the waters inside the Earth; for the Earth is hollow and has waters in its cavities. The Nile rises in summer owing to the water that comes down from the snows in Ethiopia.

The Sun, Moon and all the stars are fiery stones carried round by the rotation of the aether. Below the stars are the Sun and Moon, and also certain bodies which revolve with them, but are invisible to us. The sun surpasses the Peloponnesos [the southern half of Greece] in size. The Moon has no light of her own, but gets it from the Sun. The path of the stars goes under the Earth. The Moon is eclipsed by the Earth screening the Sun's light from it, and sometimes, too, by the [other, invisible] bodies below the Moon coming in front of it. The Sun is eclipsed at the New Moon, when the Moon screens it from us. Both the Sun and the Moon turn back in their courses [see p. 65] owing to the repulsion of the air. The Moon turns back more frequently, because it is less able to overcome the cold.

Anaxagoras also said that the Moon was made of earth, and had plains and ravines on it. The Milky Way was a reflexion of those stars that were not illuminated by the Sun. Shooting stars were sparks, as it were, which leapt out owing to the motion of the heavenly vault.

In addition, Anaxagoras put forward a general cosmological doctrine. On his theory, the creation and operation of the world were to be attributed to a Universal Mind, which he called the *Nous*. (The word is still used in English North Country dialect, to mean common sense or 'gumption'.) This doctrine was a clear ancestor of Aristotle's later teaching, according to which the rotation of the outermost sphere of the Heavens was the chief visible activity of the Divine power, or 'unmoved mover'.

All other things [taught Anaxagoras] are mixed together: *Nous* alone is unrestricted and pure and self-ruling. . . . *Nous* is the most tenuous of all things and the purest: it has universal knowledge and the greatest strength, having power over all living things, great and small. *Nous* was responsible for the whole revolution of the Cosmos, so that it started to revolve in the first place. The Cosmos began to revolve in a small way; but now the movement extends over a larger region, and it will in the future extend still further.

When *Nous* began to set this motion going, things with different properties began to separate, in proportion to the amount of the movement. And once they had got going, the rotation of the Cosmos caused them to separate even more. On the other hand, *Nous* itself pervades everything, both the things which have already separated out, and the surrounding chaos.

Dense, moist, cold and dark things came together where the Earth is now, while rare, warm, dry and bright things moved towards the distant, outer part of the aether. As these substances separate, earth is solidified, because water separates off from mists, and earth from water. Stones are solidified from earth by the cold.

FROM INGREDIENTS TO AXIOMS

For the first philosophers, then, the unchanging principles of Nature were 'underlying substances' or *ingredients*. Theories of Nature of this sort had only limited advantages. The vision they presented of all creation and annihilation as resulting from the expansion, contraction, and shuffling of unchanging material units had a certain vividness; but the picture appealed more to the imagination than to the intellect. And, since the main theme of Greek philosophy was the power of *reason*, it is not surprising that from early on people looked for alternative conceptions—in particular, conceptions lending themselves more readily to argument and rational demonstration.

So, alongside this idea of 'basic ingredients', the alternative idea grew up that *mathematical axioms* were the true principles of things. In order to explain why things are as they are and behave as they do, it is not enough (on this view) to name the material units out of which they are made: for to give such a list of ingredients alone can explain nothing with certainty. In modern physics, equally, to say only that things are made up of 'fundamental particles' explains nothing: in order to produce a genuine explanation of anything, you must suppose that these particles conform to certain fundamental mathematical equations. Explanations are *arguments*; so the bricks from which our ultimate explanations are built must be not objects, but *propositions*—not atoms but axioms.

What grounds have we for believing that such propositions can be found? Is it open to human beings to discover them? And if so, how? From Parmenides to Bertrand Russell, all these questions have been furiously debated. They lie at the heart of metaphysics and theory of knowledge, and provide an enduring link between science and philosophy.

The most important result of this passion for rational demonstration was that, in addition to theoretical physics, the Greeks invented the whole classic ideal of *abstract mathematics*. In Egypt and Mesopotamia, practical techniques of calculation had been highly developed. For example, geometry (*gē* = land, *metro* = I measure) consisted of the rules-of-thumb for use in surveying. So one finds the Babylonian mathematicians recognizing the relationship between the sides of the right-angled triangle measuring three, four, and five units; but the general theorem of Pythagoras is never stated, still less proved. Presenting mathematics as a system of general, abstract propositions, linked together by logic—like the school-book geometry of today—seems to have been a Greek innovation. Only after this could mathematics be discussed in a theoretical and detached way, apart from all practical applications.

The first great intellectual innovation of the Greeks thus led naturally on to the second: the most striking single result of the Greeks' faith that the world could be understood in terms of rational principles was the invention of abstract mathematics.

The most grandiose ambition they conceived was to explain all the properties of Nature in arithmetical terms alone. This was the aim of the Pythagoreans in Southern Italy. They knew, of course, that the phenomena in the Heavens recurred in a cyclical manner; and, when they discovered that some things on Earth also behaved in a way displaying simple numerical relationships, their ambition was reinforced. The example which made most impression on them was that of the sound emitted by a vibrating string: as they discovered, this is simply related to the length of the string. If the whole string gives out a sound of a given pitch, halving its length will immediately produce the octave, dividing it by three will produce the sound a fifth above that, and so on: concords between the original sound and its 'harmonics' always go with simple fractional lengths.

To begin with, therefore, the programme of mathematical philosophy was to look for 'the numbers in things'. And since the Pythagoreans were a religious brotherhood—for whom the natural order and the moral order were closely linked—they thought that this search would lead them to more than *explanations* alone. If one discovered the mathematical harmonies in things, one should in doing so discover how to put *oneself* in harmony with Nature. So virtues as well as sounds, shapes, and motions had to be given an arithmetical interpretation. (If this sounds strange, one must recall that the early Greeks had not been affected by later Christian views about the soul: for them, the soul was part of the natural world. A man in spiritual health was like a well-tuned musical instrument.) Yet, whatever we think of their arithmetical ethics, they had at any rate positive grounds for thinking that both astronomy and acoustics were at the bottom arithmetical; and the study of simple fractions such as we learn at school was called by the name of 'music' right on until the late Middle Ages.

How did these attempts at an arithmetical view of Nature work out in the special field of astronomy? The early Pythagoreans, like Anaxagoras, recognized that the Moon's light is borrowed, and that eclipses occur when one astronomical body obscures another. But they went further, and taught that the Earth is a sphere, rather than a disc or cylinder. Aristotle in his book *On the Heavens* gives the following general account of Pythagorean astronomy.

While most philosophers say that the Earth lies in the centre of the universe, the philosophers of Italy, the so-called Pythagoreans, assert the contrary. They say that there is Fire in the middle, and that the Earth, being one of the stars, is carried round the centre, so producing night and day. They also assume another Earth opposite to ours, which they call the Counter-Earth, but in this [Aristotle objects] they are not seeking explanations and causes for what one can observe, so much as trying to force the phenomena into the framework of their own views, and make them fit that way.

If one looks for the truth not in the observed facts but by arguing from first principles, one might well agree that the central place should not be assigned to the Earth. For the Pythagoreans consider that the worthiest place belongs to the worthiest occupant, that Fire is worthier than Earth, and that extremes are worthier than intermediate paths—the circumference and centre being extremes: from these considerations they argue that it is Fire rather than Earth which must occupy the centre of the heavenly sphere. They give as an additional reason, the need to protect the most important part of the Universe—and the centre may be so described. They accordingly call the fire which occupies this position 'the Watchtower of Zeus'.

These Pythagoreans, who say that the Earth does not occupy the centre, make it revolve in a circle around that centre, and the Counter-Earth does the same. Some people even think that there may be more bodies still revolving round the centre; only that we cannot see them because the Earth gets in the way. They give this as the reason why there are more eclipses of the Moon than of the Sun; for the Moon is obscured, not only by the Earth, but by each of the other revolving bodies as well.

There is (one might object) nothing specially arithmetical about this general picture; but the following comment by Alexander of Aphrodisias (third century A.D.) shows where the arithmetic came in.

The Pythagoreans said that the bodies in the planetary system revolve around the centre at distances related by mathematical proportions. Some revolve more quickly, others more slowly. The slower ones emit deeper sounds as they move, and the quicker ones higher sounds. These sounds depend on the ratios of the distances, which are so proportioned that the combined effect is harmonious. . . . If the distance of the Sun from the Earth (say) is twice the distance of the Moon, that of Venus three times, and that of Mercury four times, they supposed that there were arithmetical ratios in the case of the other planets as well, and that the movement of the whole heavens was harmonious. The most distant bodies (they said) move most quickly, the nearest bodies move most slowly, and the bodies in between move at speeds corresponding to the sizes of their orbits.

(Notice that in this account the planetary distances are measured from the Earth. Alexander was an Aristotelian, and so uses a geocentric example to illustrate the central point of the Pythagorean theory. The Pythagoreans would presumably have measured their distances from the centre of the whole universe.)

But if the planets emit these harmonious sounds, why can we not hear them? Aristotle reports the Pythagoreans' answer.

To meet the objection that none of us is conscious of this sound, they explain that the sound is with us right from the moment of our birth, and has therefore no contrasting silence to show it up; for silence and noise are perceived by contrast with each other, and therefore all mankind is undergoing an experience [in the case of the planets] like that of a coppersmith, who through long habit ceases to be aware of the din going on around him.

As we shall see, this Pythagorean belief that the distances of the planets from the centre of their orbits fit a simple, 'harmonious' mathematical law, was the life-long conviction of Kepler, 2000 years later, and inspired the whole course of his astronomical researches.

Notice two things about the first passage from Aristotle. To begin with: one finds in the Pythagoreans for the first time a picture of the Heavens marked at every point by 'radial symmetry'. The universe as a whole is spherical, the heavenly bodies move along perfect, circular tracks, and the Earth itself is spherical, just like the Sun, Moon, and stars. The Babylonian picture, by contrast, was more like a rectangular box; and even the early Ionians thought of the Earth as a flat disc. Yet where the Pythagoreans located the true centre of the universe is still a matter of argument. They refer to it either vaguely as 'the Central Fire', or allusively as 'the Watchtower of Zeus'. Some people—Kepler included—have argued that these phrases refer to the Sun itself, and have acclaimed the Pythagoreans as being the first to advocate a heliocentric theory. Others argue that, if they had really meant that the Sun was the centre, they would have said so more clearly. On this alternative interpretation, the Sun also moves around the Central Fire and derives its light and heat from it; by day, the whole sky over the Earth is filled with the light of the Fire; and the Sun acts simply like a lens, which concentrates this radiation more intensely in one part of the sky. If this interpretation is correct, the first serious rival to the geocentric conception of the Heavens was a view which placed the centre of the universe neither in the Earth nor in the Sun.

Notice, secondly, the *character* of the Pythagorean argument for locating the centre of the Cosmos in a Fire. Whether or not they were putting forward a heliocentric doctrine, their *reasons* for doing so were certainly not ours. What concerned them was neither techniques for working out astronomical tables, nor questions about the forces pushing the planets around—instead, they asked about the *appropriateness* of one cosmic system rather than another: 'the worthiest place belongs to the worthiest occupant'. The same essentially religious argument was to be revived by Copernicus and Kepler.

The Pythagoreans were, so far as we know, the first people to realize either the intellectual fascination—or the fun—to be got from numbers. Quite apart from astronomy and acoustics, they made a series of discoveries about the properties of whole numbers, many of which they demonstrated geometrically, by laying out pebbles in the shape of triangles, squares, and rectangles. Their sacred figure was the *tetraktys* (see page 76). This expressed for them the arithmetical equation $10 = 1 + 2 + 3 + 4$. The sum of any series of whole numbers $1 + 2 + \ldots$ (they realized) can always be displayed as an equilateral triangle by adding further rows of 5, 6 . . . pebbles to the *tetraktys*. They displayed other sums in similar ways—for instance, the sums of successive odd numbers, which form squares and the sums of successive even numbers which form oblongs (see page 76). Is it surprising if, to begin with, they assumed that similar arithmetical principles underlay all the truths and constructions of geometry?

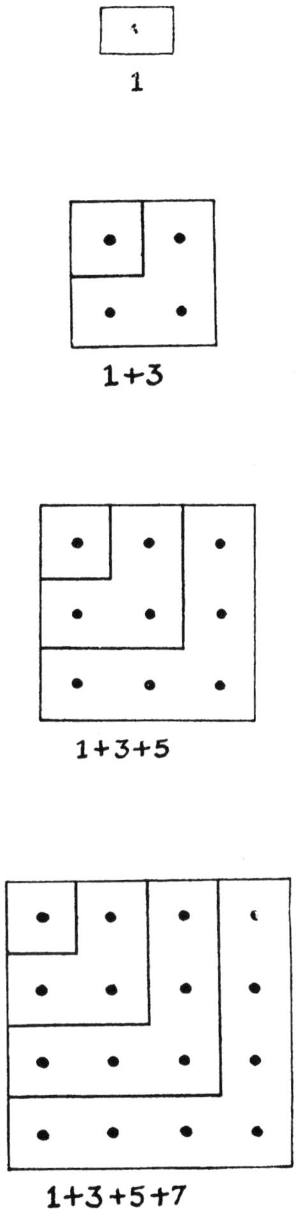

Square figures

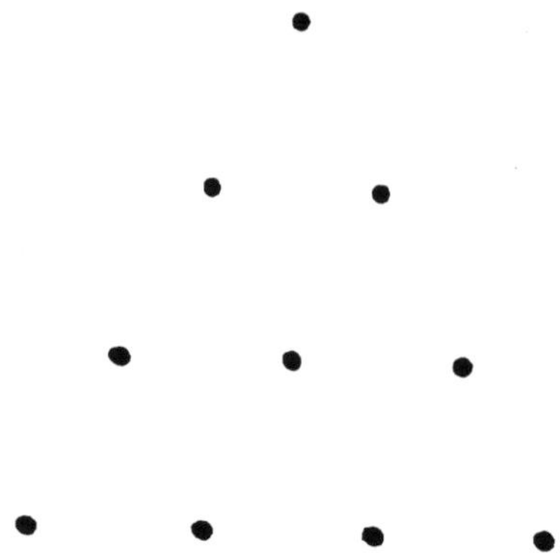

$10=1+2+3+4$

The Tetraktys

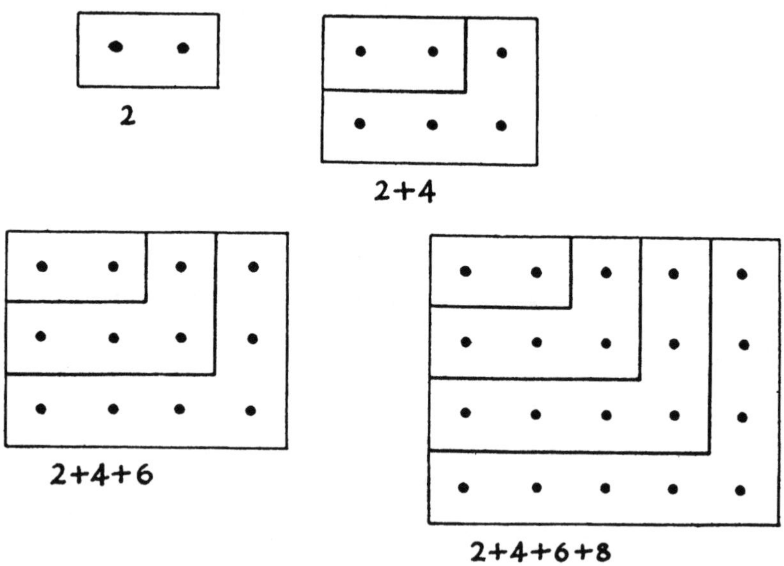

Oblong figures

The Pythagorean programme received one early set-back, and the whole direction of Greek speculation about Nature was transformed as a result. For they discovered some quite elementary geometrical relations which could not be fitted into their framework; and this came as a shock.

Consider the following problem. Suppose you have a large supply of child's building blocks *all the same length.* Now make up a square on the floor, ten bricks each way:

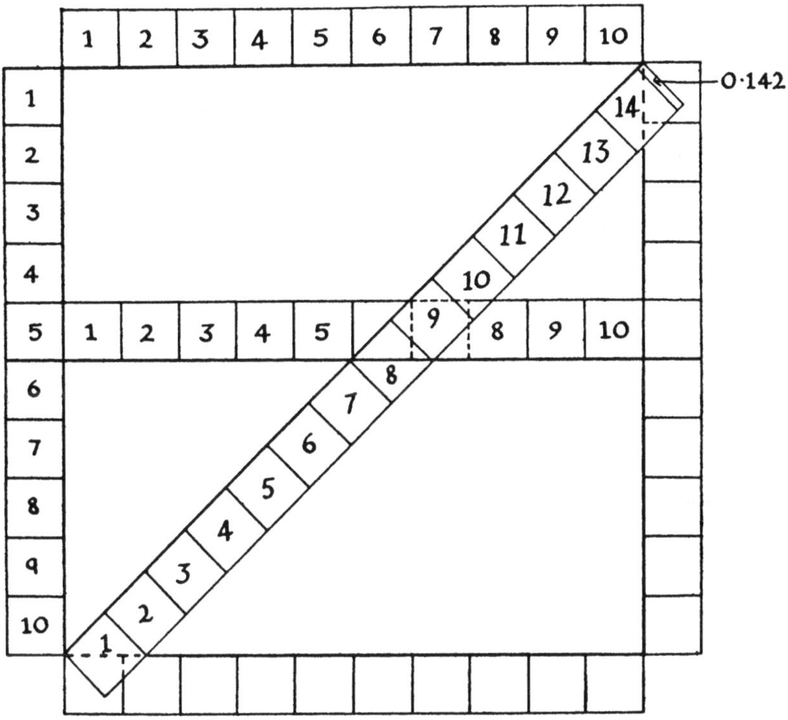

Building a square with units of equal lengths

Across the square one can place 10 blocks of the unit length exactly. *Along the diagonal* 14 blocks go in with a small length left over to the value of 0·142. . . . The total length of the diagonal can *never* be expressed as a whole number in the same units that went up to build the square

If you want to put a row of bricks across the square from the middle of one side to the middle of the opposite side, you can do this quite easily. Ten more bricks laid end to end will fit exactly, without a gap at either end. The same sort of thing can be done, however big the square is, and whatever the size of your bricks; and similar constructions are possible in a great many geometrical cases. But now try to put a row of bricks across the same square from one corner to the opposite one: when the time comes to put the last brick in place, you will not be able to fit it in exactly. Fourteen bricks will leave a gap: fifteen will be too many. You might think that this difficulty could be overcome by using small enough bricks and increasing their number sufficiently: but so long as *all* the bricks in the construction are *the same length*, the trouble will recur. However small you make your units, a part-brick will always be needed to complete the diagonal.

This difficulty can be stated in a number of ways, which amount to the same thing. You can say: a part of a brick will be needed to complete the diagonal, however small the bricks may be. Alternatively, you can say: if the side of a square is a whole number of units long, then the length of the diagonal can never be a whole

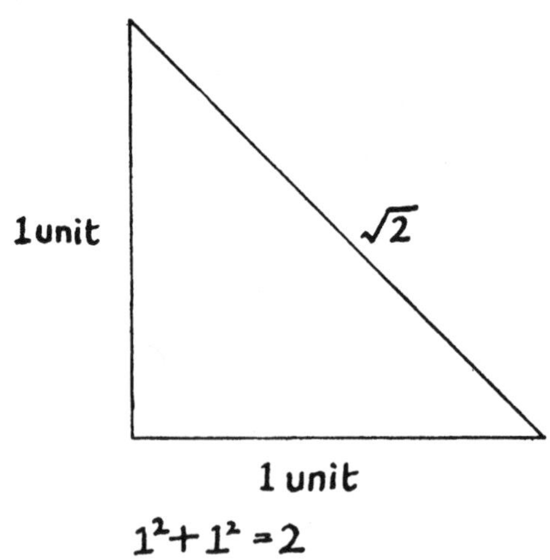

Right-angled triangles having unit sides

number of *these same units*. Or else, you can say (coining a word): the diagonal and side of a square are 'incommensurable', i.e. not-measurable-in-common-units. Or finally—since, by Pythagoras' theorem, the length of the diagonal of a square with side one unit long is the square root of two—you can say that the square root of two cannot be expressed as a simple fraction of two whole numbers. There *are* no two whole numbers which, divided one into another, equal the square root of two—and this quantity can be expressed numerically only by the endless decimal 1·4142. . . .

Mathematicians to this day still call the square root of two an *irrational* number. In doing this, they are echoing the response of the Greeks to this discovery. The whole Pythagorean view of the world was based on the idea that everything conformed to rational principles, and that these were the expression of whole numbers and their fractions (or 'ratios'); so the irrationality of the square root of two threatened to shatter their whole faith. Legend tells us that they treated the discovery as a kind of skeleton-in-the cupboard, which must be prevented from leaking out to the rest of mankind. For how could their fundamental teaching, that the integers—the whole numbers, that is—were the fundamental principles of Nature, survive the revelation that, by their standards, even simple geometry was not entirely rational?

Though their first reaction was to suppress this discovery, in the long run it had to be faced, and the results were all to the good. This set-back to an *arithmetical* theory of Nature did not (as they feared) discredit mathematics. Instead, it served as a stimulus to the production of a *geometrical* theory, which actually worked much better. After all (men thought) numbers were perhaps too general and abstract to serve as the universal principles of things: geometrical figures and models might serve physics more effectively.

PLATO'S GEOMETRICAL ASTRONOMY

The scientific programme of Plato's Academy was based on this new idea. Over the entrance to the Academy were written the words: 'Let no one enter here who knows no geometry.' It is unclear just how much geometry Plato himself knew, but at any rate he was its patron and advocate. His students made important discoveries both in pure geometry and in its applications.

It was one of Plato's followers, Euclid, who gave the first and most famous exposition of geometry as a coherent system of propositions following logically from a single set of assumptions. An earlier disciple, Theaetetus, proved a famous theorem about the five 'Platonic' solids. Suppose you take four equilateral triangles of the same size, you can assemble them together to form a pyramid (or tetrahedron); if you take six squares all of the same size, you can assemble them to form a cube; but—surprising though it may seem—only five regular solid figures, in all, can be built up from equilateral plane figures in this way. Beside the cube and the tetrahedron, these are the octahedron (made up from eight equilateral triangles), the dodecahedron (made up from twelve regular pentagons), and the icosahedron (made up from twenty equilateral triangles). Theaetetus not only showed how these five figures could be built up, but proved by a conclusive argument that there were *no other* regular solids of this kind. This result was a very striking one, which made a great impression on Plato and his whole school. The five Platonic solids reappear in Plato's theory of matter (to be discussed in another volume), and they even turn up again much later in Kepler's astronomy (*see* p. 200).

Plato retained the idea of the Pythagoreans, that our theories of Nature should rest on fundamentally mathematical principles. Ideally, he thought, these theories should take the form of rigorous arguments, whose validity was apparent to any clear-headed thinker. (In this respect, modern mathematical physics is very largely a realization of Plato's ideals.) The time was already ripe for putting plane geometry into this ideal theoretical form; and Euclid did it. But three-dimensional or 'solid' geometry (Plato complained) was neglected: its weaknesses could be overcome only by methodical research. Planetary astronomy would be the next on the list: Plato regarded this as an extension of solid geometry, though there were serious problems in the subject which had yet to be solved. The chief outstanding problem was the retrograde motion of the planets, and we shall see shortly how Plato's pupil, Eudoxos, dealt with this problem.

The one thing we find nowhere in the work of Plato or his immediate followers is a mathematical theory of *force*. The shapes of solids and of orbits could easily be discussed in mathematical terms; but the next step forward was very difficult. As Aristotle was to insist, our everyday experiences of heaviness and physical effort do not suggest that these things have anything essentially mathematical about them. None of the Greeks was, in fact, ever able to do for forces and their effects what they had so thoroughly done for numbers and shapes.

Plato discusses the purposes and methods of astronomy in the following passage from his dialogue, the *Republic*. Socrates, Plato's teacher, is depicted talking over the ideal educational syllabus with his 'feed' or 'stooge', Glaucon. They have agreed that the rulers must certainly learn a good deal of mathematics—in particular, plane geometry. Then Socrates raises the question of astronomy:

Socrates: Shall we put astronomy third? Do you agree?

Glaucon: Certainly I do. It is important for military purposes, no less than for agriculture and navigation, to be able to tell accurately the times of the month or year.

S. I am amused by your evident fear that the public will think you are recommending useless knowledge. True, it is quite hard to realize that every soul possesses an organ [the intellect] better worth saving than a thousand eyes, because it is our only means of seeing the truth; .... So you had better decide at once which party [the theorists or the practical men] you mean to reason with. Or you may ignore both and carry on the discussion chiefly for your own satisfaction....

G. I would rather go on with the conversation for its own sake in the main.

S. Well then, we must retrace our steps. We made a mistake just now about the subject that comes next after geometry. From plane geometry we went straight on to the study of solid bodies moving in circles. We ought first to take solid bodies in themselves; for next after the second dimension should come the third, and that brings us to the subject of cubes and other figures which have depth.

G. True. But this subject, Socrates, doesn't seem to have been investigated.

S. There are two reasons for that. These inquiries are difficult ones, and languish because no State thinks them worth encouraging. . . .

G. But please go on with your explanation. Just now you spoke of geometry as the study of plane surfaces, and you put astronomy next. But then you went back on that.

S. Yes, I was in too great a hurry to cover the ground—more haste, less speed. The study of solids should have come next: I passed it over because it is in such a pitiable state, and went straight on to astronomy, which studies the movements of solid objects.

G. True.

S. Then let us put astronomy fourth: assuming that the government has taken up the neglected subject [of solid geometry] and that its principles have been established.

G. Very well, then, Socrates. I will now praise astronomy on your principles, instead of vulgarly commending its utility—for which you criticized me. Anyone can see that this subject forces the mind to look upwards, away from this world of ours to higher things.

S. Anyone except me, perhaps. I do not agree.

G. Why not?

S. You interpret the phrase 'higher things' too loosely. I suppose you would think that a man who threw his head back to study the decorations on the ceiling was discovering things by the use of his reason, not his eyes. . . .

G. How, then, do you mean to reform the study of astronomy?

S. In this way. Those intricate traceries in the sky [the paths of the stars and planets] are, no doubt, the loveliest and most perfect of *material* things, but they are still part of the visible world, and therefore fall far short of the true realities—the true movements, in the *ideal* world of numbers and geometrical figures which are responsible for these rotations. Those [theoretical principles], you will agree, have to be worked out by reason and thought, and cannot be observed.

G. Exactly.

S. Accordingly, we must use the embroidered heaven as an example to illustrate our theories—just as one might use exquisite diagrams drawn by some fine artist such as Daedalus. An expert in geometry, faced with such designs, would admire their finish and craftsmanship; but he would not dream of studying them in all earnest, expecting to find all angles and lengths conforming exactly to the theoretical values.

G. That would of course be absurd.

S. The genuine astronomer, then, will adopt the same outlook when studying the motions of the planets. He will admit that the sky and all it contains have been framed by their maker as perfectly as such things can be. But when he considers the proportions of day to night; or of day-and-night to month; or of month to year; or the relative periods of the different planets to the Sun and Moon and to each another—he will not imagine that these visible, material changes go on for ever without the slightest alteration or irregularity, and waste his efforts trying to find perfect exactitude in them.

G. Now you put it like that, I agree.

S. So if we mean to study astronomy in a way which makes proper use of the soul's inborn intellect, we shall proceed as we do in geometry—working at mathematical problems—and not waste time observing the heavens.

Plato is here so very insistent that astronomers must concentrate on theoretical problems, and ignore the 'technology' of their subject—navigation and the calendar—that he belittles the role of observation. Many modern readers find this attitude wholly objectionable: observation, they insist, is the life-blood of science. Yet we should not be put off by what he says. For, in the first place, in Plato's time a great deal more was known about the observable motions of the planets than any theory yet explained. Undoubtedly what astronomy then needed *was* some hard slogging on the theoretical side, rather than yet more observations. Secondly, Plato was concerned with the educational value of astronomy—which he saw as a training for the intellect. Problems of mathematical astronomy (he considered) would be just as good exercises in reasoning as arithmetical or geometrical ones.

But the crucial point is the following. A theory of the Heavens must not just describe how the stars and planets appear to move: it must *make sense* of those movements. We shall understand Nature (Plato thought) only if we can get past the day-to-day changes we *see*, to underlying ideals with which we can be *intellectually* content. Uniform motion in a perfect circle was, to him, one of these ideals: it was a kind of motion which he found quite self-explanatory. Why was this particular ideal so attractive? Here logic and observation probably both played a part. Mathematical truths are unchanging: if they are acceptable today, they are acceptable for all time. A body moving in a circle can go on doing so for ever, since its path never reaches an end; and since all the radii of a circle are of the same length, its distance from the centre of its orbit will also be unchanging. With this ideal in mind, the one thing about the Heavens Plato found completely intelligible was the apparent motion of the fixed stars. No material things, he said, could more perfectly demonstrate the 'eternal truths' of geometry.

The more intricate paths of the Sun, Moon, and planets, on the other hand, are not immediately intelligible in these terms. Yet they must be made so. Presumably the same principles govern all celestial movements: the fundamental problem for astronomical theory, therefore, took the form, 'How can we construct geometrically tracks such as those of the planets, using only concentric circular movements as our basic idea?'

When Plato wrote the *Republic*, this problem had not been solved. But he was already able to indicate in outline the sort of picture of the cosmos that his principles demanded. The picture is presented through the mouth of Socrates as part of the Myth of Er—a brief allegory which represents a kind of distant ancestor to John Bunyan's *Pilgrim's Progress*. Plato depicts the cosmos as a complicated spinning-top, built up out of eight shells, all turning independently on a common shaft. The shells move at slightly different speeds: the seven inner ones, carrying the planets, lag slightly behind the outermost shell, which carries the fixed stars. (Plate 4.) Geometrically, Plato's model gave a satisfactory first account of the motions of the heavens. But when it came to explaining what caused the movements of the shells—what agencies were involved—a theory of this type could give no answer. Allegorical figures alone are invoked—Necessity and her three daughters, The Fates: we are back again in the mythological realm of personified natural causes. All the same, a crucial step forward has been taken.

Now when each company had spent seven days in the Meadow, on the eighth they had to rise up and journey on. And the fourth day afterwards they came to a place whence they could see a straight shaft of light, like a pillar, stretching down through the whole of Heaven and Earth—more like the rainbow than anything else, but brighter and purer. They came to this pillar after a day's journey, and there in the middle of the light they saw the ends of its chains stretching down from Heaven; for this shaft of light binds the heavens, holding the revolving firmament together, like the girths under the belly of a ship.

And from the ends stretched the Axle of Necessity, on which all the circles revolve. The shaft and hook were adamantine, and the body was made up as follows. In shape it was like an ordinary spinning-top; but from Er's account we must picture it as consisting of a large top, having the inside completely scooped out, and inside it a second smaller top, and a third and a fourth and four more, fitting into one another like a nest of bowls. There were in all eight tops, set one within another, their rims forming (as it were) the continuous upper surface of a single top and showing from above as circles around the axle, which pierced right through the centre of the eighth top. The circle forming the rim of the first and outermost whorl [Fixed Stars] is the broadest; next in breadth is the sixth [Venus]; then the fourth [Mars]; then the eighth [Moon]; then the seventh [Sun]; then the fifth [Mercury]; then the third [Jupiter]; and the second [Saturn] is narrowest of all. The rim of the largest whorl [Fixed Stars] was spangled; the seventh [Sun] brightest; the eighth [Moon] coloured by light reflected from the seventh; the second and fifth [Saturn, Mercury] like each other and yellower; the third [Jupiter] whitest; the fourth [Mars] somewhat ruddy; the sixth [Venus] second in whiteness. The axle revolved as a whole in one direction; but, as it turned, the seven inner circles revolved slowly in the opposite direction. Of these the eighth [Moon] moved most swiftly; the seventh, sixth, and fifth [Sun, Venus, Mercury] were second in speed and moved in conjunction; next in speed was the fourth [Mars], giving the appearance of turning back on itself; next came the third [Jupiter]; and the slowest of all was the second [Saturn].

The axle turned on the knees of Necessity. Upon each of its circles was a Siren, who was carried round with its movement and gave out a single sound of constant pitch, so that all eight between them made up a single scale. Round about, at equal distances, the three daughters of Necessity—the Fates—were seated on thrones, robed in white with garlands on their heads. These were Lachesis, Clotho, and Atropos, and they chanted to the music of the Sirens: Lachesis of things past, Clotho of the present, and Atropos of things to come. From time to time Clotho laid her right hand on the outer rim of the axle and helped to turn it, while Atropos likewise turned the inner circles with her left hand and Lachesis took hold alternately of the inner and outer circles with either hand.

Compare this passage with the one quoted earlier from Vitruvius to illustrate the Babylonians' knowledge of planetary movements. Vitruvius simply reports the way in which the heavenly bodies are *seen* to move across the sky, as viewed by an observer standing on the Earth. Anyone who can name and recognize the different heavenly bodies can understand what he says. Plato, by contrast, introduces a new, theoretical element: to understand him, we must not just look at the sky, but must think about the Heavens—using our imaginations. Plato's principles being what they are, the intellectual model he offers us is a geometrical one. Instead of staying on the Earth and reporting what we see as we watch the sky, we must now imagine ourselves outside the whole universe, and ask ourselves what its structure must be in order to give rise to the familiar happenings visible in the sky. His picture of the cosmos as a nest of eight concentric shells around the Earth is the result.

Plato realized that his vision of the cosmic system was not only sketchy, but contained one serious loose end. These faults must be remedied, if one was really to understand the principles behind the celestial motions. All he had produced was a general outline picture, which had yet to be worked out in detail: radii and speeds had to be worked out for the various shells. And one phenomenon actually appeared irreconcilable with Plato's general scheme. This was the 'retrograde motion' of the planets, which we have already encountered in the Babylonian records. (Notice the reference in the last quotation to the way in which the fourth sphere—that of Mars—gave the appearance of turning back on itself.) The problem was to account for this looped track without abandoning his fundamental theoretical model, of steady, regular, and continuous circular motion.

It is not surprising that Plato set his pupils in the Academy the task of finding a geometrical construction by which this phenomenon could be brought into the general scheme. The man who came nearest to a satisfactory solution was Eudoxos of Knidos. His model of the planetary system required, not eight, but twenty-seven spheres: three each for the Sun and Moon, four for each of the five known planets, and one for the fixed stars. The twenty-seven spheres spun around a common centre—the Earth. Each sphere turned at a uniform speed on its own axis, but the system was so constructed that one sphere could share in the movement of neighbouring spheres through the attachment of its axis. The final retrogradation was a consequence of two things: (*a*) this superimposition of movements of three or four spheres, and (*b*) the fact that the axes of the spheres were not all in the same plane.

Each of the four spheres required for any one planet had its own functions. The outermost of the set of four spheres accounted for that part of the planet's movement which it shared with the whole firmament of stars—rising and setting once every twenty-four hours. The second sphere gave the planet its movement along the ecliptic, carrying it right around the Zodiac in a period ranging from a month for the Moon to thirty years for Saturn. The two remaining spheres, whose axes were set at an angle to the others, accounted for variations in speed. The planet itself was carried by the fourth, innermost sphere, whose observed motion was the resultant of all four.

This system is very difficult to visualize even with the aid of a diagram; but it worked, producing the peculiar looped track of the planets out of uniform circular motions alone. As an intellectual construction, the whole system is very ingenious: exactly how it works out is too complex to explain in detail here. The way Eudoxos produces the 'loop', by combining the motions of the inner two spheres, is particularly elegant.

Eudoxos, then, had attempted to make coherent mathematical sense of the Heavens in a new way. Yet just how much had he achieved? And which way forward was astronomy to progress after him?

Certainly his theory—as it stood—went only part-way towards giving us a convincing picture of how the planetary system *works*. All Plato had asked for, and all Eudoxos had so far provided, was an intellectual construction capable of fitting the chief planetary phenomena into the general geometrical framework. From Plato's point of view it was irrelevant to ask whether the twenty-seven spheres were real material things: they were mathematical ideals, not solid bodies. Nor had the full implications of the new model been checked in detail, to see just how life-like a picture it yielded of the actual pageant of the Heavens.

At this point, the road forked. One could concentrate on fitting the theory to the facts: adding to the basic system as many further geometrical sophistications as were needed to match the observed motions. (This line of development we shall take up again with Ptolemy.) Alternatively, one could take the success of Eudoxos' geometrical system as a starting-point, and follow up the further questions it raised. Assuming that the twenty-seven spheres had some genuine physical reality, how did they interact? What forces were needed to keep them all moving as they did? What were the spheres made of, and did celestial bodies move according to the same principles as terrestrial ones? One could go on (in other words) from mathematics into physics.

In the subsequent history of Greek astronomy, these two demands pulled in different directions. The desire to make the geometrical model physically intelligible often conflicted with the need to make it conform more exactly to the facts. To begin with, physics pulled more strongly. There is some evidence that even Eudoxos himself was drawn in this latter direction—that he was in spirit a mathematical physicist rather than a mathematician alone.

He seems to have wondered whether material counterparts really existed up in the sky corresponding to the mathematical spheres of his construction—whether the spheres were not so much imaginary as *invisible*. It is almost as if he had been groping towards the more recent form of physical theory, in which mathematical equations are not just intellectual ideals, but are taken to refer to some mechanism that really exists, even though we cannot see it or touch it. (A case in point is Rutherford's picture of the atom as a miniature solar system.)

For Plato himself, Eudoxos' system of spheres was a pure mathematical construction. Eudoxos himself was already feeling his way beyond pure geometry. And Aristotle was soon to turn the whole system into a solid, mechanical train of gears.

FURTHER READING AND REFERENCES

The transition from the mythological to the rationalistic phase in Greek thought is discussed in

F. M. Cornford: *From Religion to Philosophy*
W. K. C. Guthrie: *The Greeks and their Gods*

The standard general account of the early Greek natural philosophers (and still in many ways the most useful) is

J. Burnet: *Early Greek Philosophy*

There is a particularly good discussion of the Pythagorean views in

S. Sambursky: *The Physical World of the Greeks*

For a detailed exposition of the astronomical theories of the early Greeks, see

T. L. Heath: *Aristarchus of Samos*
J. L. E. Dreyer: *A History of Astronomy from Thales to Kepler*

For a brief popular treatment of the subject see

B. Farrington: *Science in Antiquity*

The bearing of astronomical problems on Plato's philosophical system is discussed in

F. M. Cornford: *Plato's Cosmology*

3

The Premature Synthesis

THE progress of science always involves a delicate balance between critical observation and speculative theorizing—between careful, piecemeal investigation of particular problems and imaginative, general interpretation of the results obtained. Individual pieces of research work may, in themselves, be pedestrian. Why (we might ask) should anybody spend day after day watching bees walk about on the honeycomb? And why should anybody spend money building polythene balloons to carry photographic plates into the upper atmosphere? How much more striking (one might feel) to hit the Moon with a space rocket! Yet research topics will appear to the outsider quaint or boring only if he fails to recognize their wider implications: they are worth investigating just *because* of their power to throw light on more general issues. One studies the 'dances of the bees' at length not out of mere inquisitiveness, but because here, for once, we find creatures in the animal world passing on information to each other almost as though they had a language. And one sends packages of raw film high into the stratosphere so as to expose them to high-energy cosmic rays, whose effects on the photographic emulsion have a bearing on our theories of nuclear structure.

Space-rocketry, on the other hand, is not really a scientific activity at all, so much as a *technological* one. The possibility of designing space rockets may depend on new scientific discoveries—for instance, on our knowledge of quick-burning fuels—and having the rockets will put us in a position to find out directly about conditions in the inter-planetary regions. But, taken by itself, learning how to dispatch a projectile to the Moon throws light on no new natural facts or general principles; so it is not *science*. For in science the theoretical conclusions are what count. Other things being equal, the wider the implications of an investigation, the more important it is. And the more general the conclusions on which a man is prepared to stake his reputation, the more hostages he gives to fortune—knowing that the adequacy of his ideas will finally have to be judged by his successors in later generations.

There are two kinds of advance in science which, when they come off, are particularly striking. First: when several different branches of a science have developed along independent lines, people justifiably try to bring them into harmony within the wider framework of a more general theory—for example, uniting the fundamental theories of electricity, magnetism, and optics, as Maxwell did. As a matter of method, the ambition to establish such connections between different branches of a science is wholly legitimate. Secondly, some scientific advances may have repercussions even outside the boundaries of science, producing radical changes in the ordinary man's view of the world. Thus, Newton demonstrated that comets are, as a matter of dynamics, no different from planets; and this implied that they need not be regarded as visitations or portents. So people may quite rightly ask how far the results of science support or discredit popular ideas about the world in which we all live. Newton's theory of motion and gravitation was important in both these ways: it successfully pulled together into a single scheme the results of half a dozen independent lines of investigation, and it did so in a way which had a profound impact on 'common sense'. Two thousand years earlier, Aristotle had had the same ambition.

ARISTOTLE'S PROGRAMME

To understand the direction which Greek physics took after Eudoxos, one needs to bear this in mind. Babylonian astronomy was essentially piecemeal, and devoid of general theory. The early Ionians and Pythagoreans had been ready enough to speculate in a general way, but they never succeeded in giving their ideas any substantial foundation in fact. Eudoxos' geometrical spheres at last seemed to offer a really solid basis for a theory about the mechanism of the Heavens; and the opportunity he provided was quickly taken up.

Such a theory had to be built up in two stages. First, one had to reach some general understanding of the ways things move, by studying motion close at hand on the Earth; after that, Eudoxos' system could be the clue, or stepping-stone, for extending one's theories to the Heavens. By comparing the celestial mechanism with terrestrial ones, one could then confirm or refute popular ideas about the Heavens and the Earth. This was the programme Aristotle set himself in his work on dynamics and astronomy: the results are set out in his treatises *On Physics* and *On the Heavens*.

Aristotle was fired by the ambition to unify all the separate branches of natural philosophy—and to show their bearings on natural theology. His work formed the first great synthesis in science. He so nearly achieved complete success that men took almost 2000 years to build a better system of physics which covered as wide a ground. In the long run his system broke down. But this fact is no reflection on its creator. His programme was a legitimate one; his conclusions were—so far as they went—very largely correct; and the synthesis he attempted in physics was eventually completed by men who profited from his experience.

It has been fashionable to pour scorn on Aristotle's physics. Yet one need not lean over backwards in order to do him justice. His general principles may have been superseded, yet many of his most-criticized results are sound enough, if looked at in the context in which he presented them. Some of them, indeed, have retained a respected, though subordinate, place in more recent science. In dynamics, his starting-point appears to modern physicists hopelessly ill-chosen. But, in his time, the mathematical analysis of motion had not been carried far enough for anyone to set dynamics on its modern basis. And when he applied his dynamical principles to astronomy, the result was a picture which in the last few centuries men have entirely dismantled and abandoned. Yet it was natural enough for him to study first how bodies moved on the Earth, and to argue from this to conclusions about the Heavens: he could hardly have foreseen that the reverse would prove more fruitful.

He might have played for safety, and pursued to an even more minute level the methods of geometrical analysis developed by Eudoxos. If he had done so, he might have been a greater mathematician, but he would certainly have been a lesser scientist. For in science, the courage to generalize is a major virtue; and one cannot judge a man's stature solely in terms of his long-run success.

So it is not enough to sweep aside Aristotle's ideas about physics as intellectual blunders. Unless we take the trouble to reconstruct his arguments with some care, we shall neither see what needed doing to prepare the ground for Galileo and Newton, nor be able to understand their true achievements. Still less shall we recognize why Aristotle's ideas held the field for so long.

In fact, the case Aristotle put forward was very strong. He analysed intelligently a wide range of careful observations. His conclusions fitted admirably with our common experience of the world, and in many respects further experiment and observation would have reinforced his ideas rather than refuting them. Far from being a piece of armchair imagining, his theory of motion was if anything too close to the facts and not abstract enough. By contrast, when Galileo and Newton stood back and looked at the facts afresh from the safe distance of a mathematician's study, they were led to re-shape dynamical theory in a form which affronted common experience.

MOTION AND CHANGE

We must now look directly at Aristotle's dynamical ideas, and see in the succeeding section how he applied them to astronomy. To begin with, these ideas must be related to his overall intellectual goal.

Aristotle, like Eudoxos, was a pupil of Plato at the Academy; but he never entirely shared the geometrical ideal of a scientific explanation. By temperament, he was not a mathematician, and his differences with Plato's successors became so marked that he left the Academy and set up on his own. Of course, he shared the ultimate ambition of all Greek philosophers, that one should find the general principles of Nature; but his own personal interests and experience lay in a direction so completely different from Plato's that, over the years, he built up quite a different ideal for science.

Aristotle's father was a doctor and acted as personal physician to Philip of Macedonia. He himself was a brilliant zoologist, with a particular flair for marine biology—indeed, some of his discoveries were thought for many years to be old wives' tales, and have only recently been 'rediscovered'. This preoccupation with zoology kept the complexity, variety, and vitality of Nature well in the front of his mind. As a result, he was never convinced that one either could or should reduce the workings of Nature to abstract, mathematical terms.

He allowed mathematics a certain importance and value, but considered that its scope was limited. He could understand treating numbers mathematically; and also distances and times—though comparatively, rather than absolutely. One distance could be $\frac{1}{2} \times$ another, and one period of time $2 \times$ another, but all measurements (he saw) involved comparison. A six-foot man has a height $6 \times$ the length of a foot-rule, but only $2 \times$ the length of a meter rule. His height, therefore, has nothing intrinsically '6' or '2' about it. Yet in Aristotle's opinion many properties of things could not be understood in geometrical or numerical terms, even to this extent. We now take it for granted that the difference between heavy things and light ones can be treated mathematically, in terms of their 'quantity of matter' or 'mass'. But in Aristotle's view heaviness and lightness were sensory qualities, no more numerical than a sweet taste or a bad smell. Even speed led to difficulties: you cannot 'divide' a length by a time and still get a *pure* 'ratio'.

A mathematical theory of natural changes was (he thought) defective in two ways: it was too abstract, and it covered too little. An adequate theory must treat changes of *all* sorts as equally authentic and significant—changes which could be expressed numerically being only one of many possible sorts of change. His idea of change was completely general; alterations in colour, shape, health, or mental state all being different varieties, each with its own characteristics. For example, how could one give a complete account of the growth and ripening of an apple in terms of numbers and shapes alone?

Starting as a biologist, Aristotle recognized two main types of alteration which can affect a body. On the one hand, it can change through its own natural sequence of growth and development: a seed develops first into a seedling, then into a grown plant, then flowers, sets fresh seed, which in its turn develops. . . . This is 'natural' change. On the other hand, some alterations take place because an outside body has intervened, producing changes which have nothing to do with the normal sequence of development: e.g. a plant may be trodden on, or pruned, or clipped into some unnatural shape. Change of this second sort is 'forced' change.

The distinction between forced and natural changes is important, because (as Aristotle saw) nearly all the properties of bodies can be altered in either of these ways, and the two sorts of change call for quite different explanations. An apple can redden either because it has ripened naturally, or because a maggot has upset its metabolism, or because a small boy has painted it for Hallowe'en. To explain a natural change, we need only demonstrate that it was characteristic of the 'species'—whether animal, vegetable, or mineral. To explain forced changes, on the other hand, one must point to the outside agents responsible. At the everyday level of interpretation, this distinction between things that 'happen naturally' and things that 'are made to happen' is a reasonable one.

Having drawn this distinction completely generally, Aristotle brought his theory of motion—i.e. change of *position*—into line. Motion, too, must be either natural or forced: things either move of their own accord, or they are forced to move by an outside agent. We can see both kinds of motion on the Earth. Rocks and streams, left to themselves, roll or flow down a hillside until they find their natural level, when they stop: they will go uphill only if carried. The flames and smoke from a fire rise vertically upward through still air as high as we can see: they, too, Aristotle presumed, would cease moving only at their natural level, somewhere in the upper atmosphere, and they would stop short of this only if forcibly held down. In each case, the natural motion was vertical, and limited in duration: a body on the Earth could move sideways, or keep moving indefinitely, only if it was pushed or pulled. And this seemed to bear out one of Aristotle's general ideas about terrestrial change. Impermanence was the mark of earthly things. Motion—growth—even human life itself ran its course, and came to its natural end.

We have only to look at the Heavens to see a complete contrast. The Sun, the Moon, and the fixed stars circle around us continuously, and nothing appears to be pushing them. This unceasing movement, Aristotle argued, must be the natural motion to expect of celestial objects. In all respects they are as permanent as earthly ones are transitory: their light never fails, they reappear with complete punctuality, there is—so far as our personal experiences go to show—nothing impermanent about them. The circle, likewise, provides an unending and undeviating path, symmetrical on all sides and having neither end nor beginning. To put the argument in his own words:

> From the Heavens other things derive their existence and life. . . . So the popular idea about divine, celestial things is that, being primary and supreme, they are necessarily unchanging. This confirms what we have already said. For there is nothing else stronger than the Heavenly Sphere that could make it alter—since if there were, it would have to be *more* perfect. Whereas the Heavens have no defects, and are all they need to be. So the unceasing movement of the Heavens is perfectly understandable: everything ceases to move when it comes to its natural destination, but for the body whose natural path is a circle, every destination is a fresh starting-point.

Circular motion was therefore not only what one *saw* in the Heavens, but also what theory would lead one to expect—and even what theology required. Physics and astronomy seemed in this way to confirm what ordinary men had for so long taken for granted: the absolute contrast between the Divine Heaven and the mortal Earth.

Before considering the astronomical consequences of this doctrine, we must look in detail at Aristotle's analysis of terrestrial motion. We must see, first, how he thought the speed of a body was related to the force pushing it; second, how the speed was related to the resistance opposing its motion; and lastly, why he concluded that a complete vacuum could not exist in Nature.

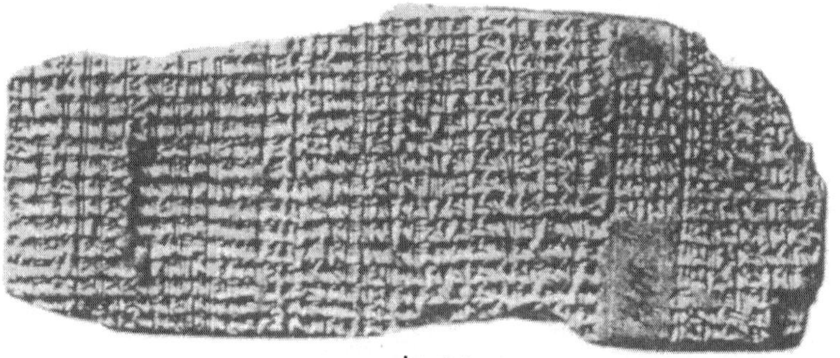

obverse

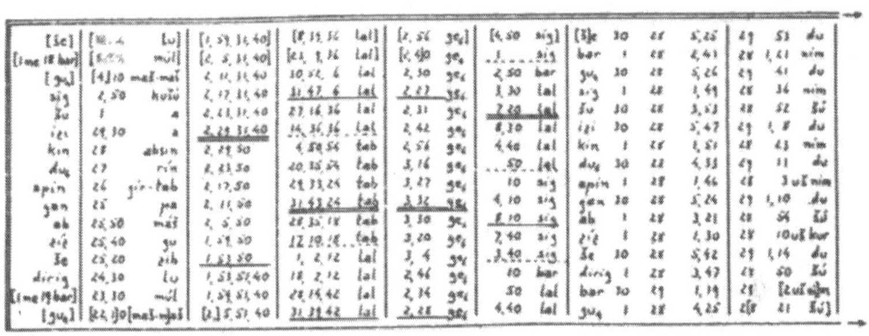

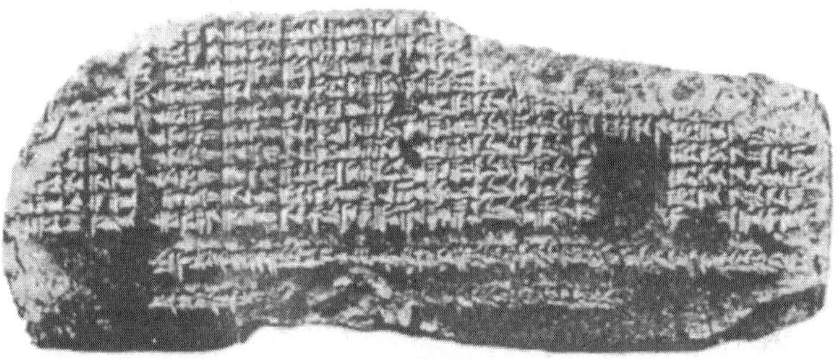

reverse

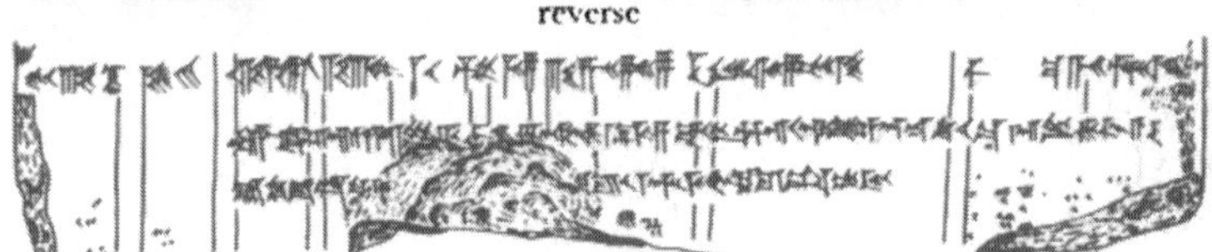

VENUS, 1960

FOR 0^h EPHEMERIS TIME

Date		Apparent Right Ascension		Apparent Declination		Semi-diam-eter	Hor. Par.	True Distance from the Earth		Ephem-eris Transit
		h m s	s	° ′ ″	″	″	″			h m s
Apr.	1	23 24 01·95		− 5 24 48·3		5·40	5·66	1·556 051		10 46 54
			+274·21		+1692·6				+ 4 051	
	2	23 28 36·16		4 56 35·7		5·39	5·64	·560 102		10 47 32
			273·74		1701·1				4 012	
	3	23 33 09·90		4 28 14·6		5·38	5·63	·564 114		10 48 09
			273·32		1708·8				3 974	
	4	23 37 43·22		3 59 45·8		5·36	5·61	·568 088		10 48 45
			272·93		1716·0				3 936	
	5	23 42 16·15		3 31 09·8		5·35	5·60	·572 024		10 49 21
			272·57		1722·2				3 898	

PLATE 1. Babylonian astronomical tablet in cuneiform script with excerpt from *Astronomical Ephemeris* for comparison

PLATE 2. Main stairway of the Ziggurat as excavated at Ur

PLATE 3. The port of Pythagoreon on the island of Samos, the reputed birthplace of Pythagoras

PLATE 4. Sixteenth-century brass armillary sphere: note the circles for the Sun and Moon and the central axis through the poles, and compare with the description of the planetary system in Plato's *Myth of Er*

PLATE 5. Illuminated Arabic text illustrating persistence of star-worship in Mesopotamia more than a thousand years after Babylonian times

a

b

c

PLATE 6. Stellar parallax: objects on the Earth show a different pattern when viewed from different angles. (Compare (*a*), (*b*) and (*c*).) The constellations show no such change from season to season. Seen through a telescope, the Constellation of Orion contains many stars invisible to the naked eye—as Galileo first showed in his drawing (*d*) of the Belt and the Sword

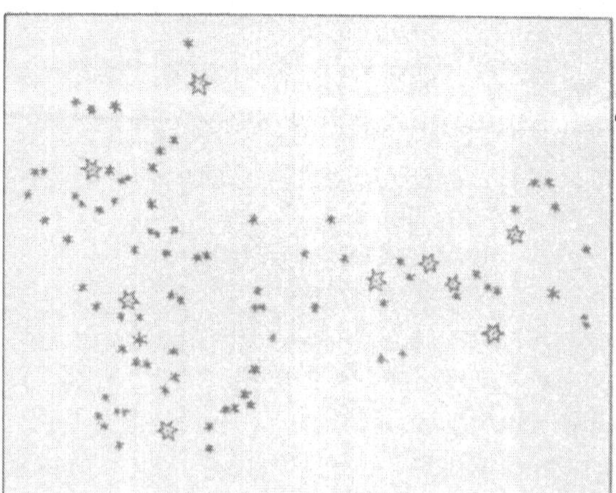

d

PLATE 7. To mediaeval man, comets were mysterious visitants whose place in the sky was not understood; this picture is taken from the Bayeux tapestry

PLATE 8. The *Book of Hours* of the Duc de Berry displayed phases of the Moon and other astronomical data in addition to a normal calendar

PLATE 9. The Ptolemaic world system, from *Cosmographia*, Apianus, 1553

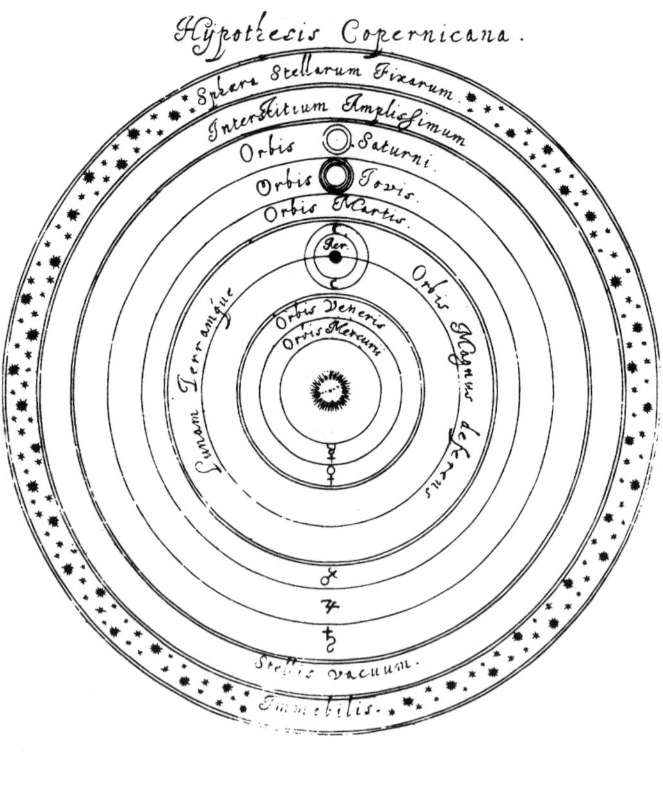

PLATE 10. The Copernican world system, from *Selenographia*, J. Hevelius, 1647

PLATE 11. Tycho Brahe devised large, highly accurate direction finders for use in his observatory at Uraniborg in Denmark

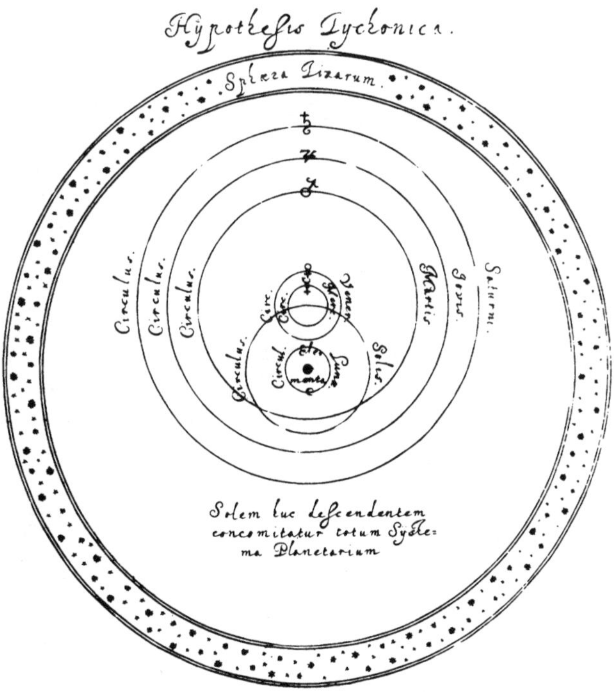

PLATE 12. The Tychonic world system, from *Selenographia*, J. Hevelius, 1647

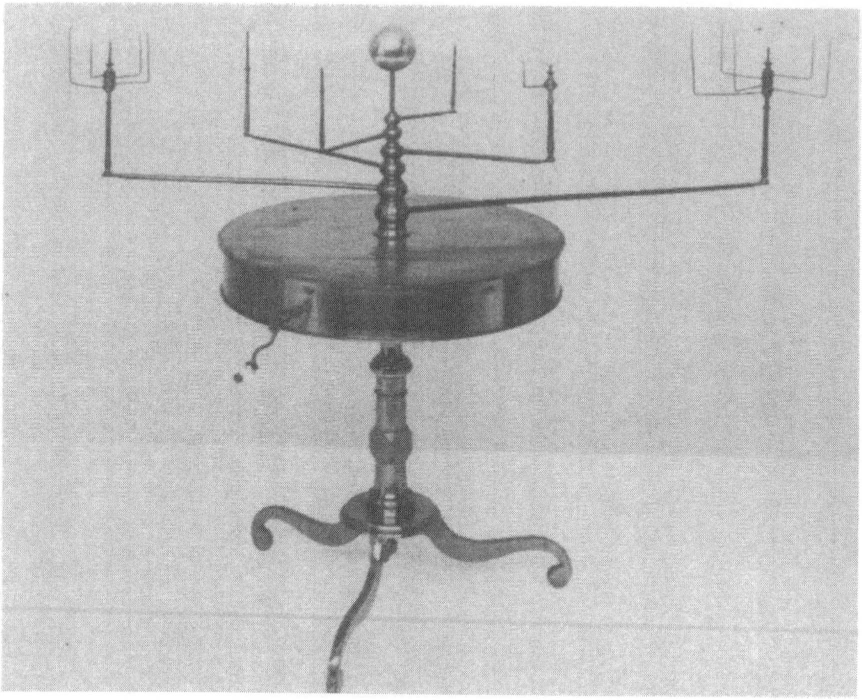

Plate 13. The new picture of the solar system was shown in models (called orreries) which were widely made during the eighteenth century

Plate 14. These two spiral nebulae, one seen end on, the other sideways on, are believed to resemble the galaxy of which our own solar system is a part

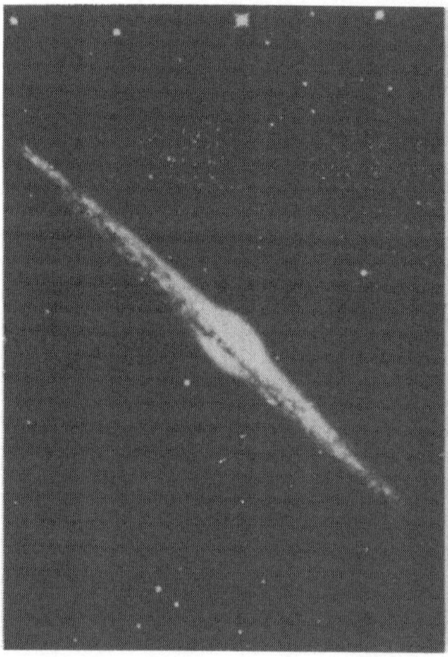

a

b

(i) *Speed and Effort*

At the very base of Aristotle's theory of dynamics is a doctrine which has often been misinterpreted. For the first principle of his theory *looks* very like the assertion:

$$\text{Force} = \text{Mass} \times \text{Velocity}$$

i.e. a given force will move a body with a speed proportional to the force and inversely proportional to the mass of the body. Yet modern physicists would condemn such a doctrine as entirely fallacious. The true relation—they would say—was first clearly formulated by Newton:

$$\text{Force} = \text{Mass} \times \text{Acceleration}$$

i.e. a given force will produce in a body, not a *speed*, but a *change in speed* proportional to the force and inversely proportional to the mass of the body.

If we interpret Aristotle's dynamical theory in this way, it will be clearly wrong; and we shall inevitably be prejudiced against it from the start. But this is not a fair interpretation: the words force, mass, and velocity acquired their present theoretical meanings only in the seventeenth century, so to use them in stating this view is to invite misunderstanding. Aristotle, in fact, never put forward any general formula of the modern kind for the moment-to-moment relations between such mathematical variables as force, mass, distance, and time. The questions he was concerned with were much more down-to-earth: e.g. 'How long will it take a given team of men to shift a given body a given distance?' Mathematically, his theory never went beyond proportion-sums of the Boys and Buns variety—'If three boys eat twelve buns in four minutes, how long will eighteen boys take to eat twenty-seven buns?'

His general conclusions about motion were stated in these very terms: laying down 'rules of proportion' and specifying their scope.

Suppose, then, that a mover has shifted a body a given distance in a given time, then in the same time the same effort will shift half the body twice the distance, and in half the time it will shift half the body the whole given distance: in this way the rules of proportion will be observed. . . . But it does not necessarily follow that the same effort will shift a body twice the size half the given distance in the same time. If it were purely a matter of ratios, one man would be able to move a ship single-handed—since the combined efforts of a whole team of ship-haulers and the distance they can move the ship could both be divided by the number of men in the team. This refutes Zeno's argument that a single grain of corn *must* make a noise when it falls: for—however long it acts—so small a part may by itself entirely fail to move that air which a whole bushel of corn will move when it falls.

(Notice how here Aristotle immediately extends his generalization about motion to cover a conclusion about noise: as we saw earlier, all kinds of change are, for him, different varieties of a single general kind.)

(ii) *Speed and Resistance*

The second of Aristotle's general principles concerns the effects of resistances to motion. Suppose his general question is: 'How long will it take, by the exercise of a given effort, to shift a given body a given distance?' As he sees, the right answer to this question is: 'It depends what resistances have to be overcome.' If his ship-haulers pull a ship across a greased surface, they will clearly cover the ground much more quickly than if they pull it over a stony track. And if you wade with sea-water up to your knees, especially against the current, you will take much longer to go a hundred yards than if you walk the same distance along a beach.

The medium causes a difference [in the motion] because it impedes the moving body, most markedly if it is moving in the opposite direction, but to a lesser degree even if it is at rest; and this is particularly true of a medium that is not easily cut through, i.e. a medium that is on the thick side. A body will move through a given medium in a given time, and through the same distance in a thinner medium in a shorter time, in proportion to the thicknesses of the hindering media. Consider water and air: it will move through air faster than through water by just so much as air is thinner and less corporeal than water. . . . So generally: the less resistant and the more incorporeal and easily-divided the medium, the faster in proportion will be the movement.

(What Aristotle here means by thickness is what we might call 'gooeyness'—or, to use the technical term, viscosity: this is related to, but not identical with, the density or mass-per-unit-volume of the two media.)

Accordingly, Aristotle thought of all movement as a balance between effort and resistance. A certain initial effort was required for a body to be shifted at all; and the rate at which it would then cover the ground was determined by the balance between effort and resistance. Whenever he theorized about a moving body he compared it for purposes of argument with a standard object being shifted at a steady rate against a uniform resistance. This was his 'explanatory paradigm'. Here lies the crucial difference between his system of dynamics and the modern one: it is not that they say contradictory things about the relation between force, velocity, and acceleration, but that they have quite different paradigms—they place quite different examples in the centre of their theoretical pictures. Aristotle's example is that of a body moving *against a constant resistance*, Newton's is that of a body moving *in the absence of any resistance*. So no wonder their first principles look so different.

Combining Aristotle's first two rules of proportion, one arrives at the following result. The rate at which a body moves is in proportion to the effort exerted on it, and inversely proportional both to its bulk and to the resistance opposing its motion. If we state this doctrine in a precise, mathematical form—though this would be foreign to his way of thinking—we get the formula known in modern physics as Stokes' Law. This states that the speed with which a body moves through a viscous medium is proportional to the force acting on it, and inversely proportional to the viscosity. This is a well-established principle of hydrodynamics, and holds accurately for motion through any fluid of sufficient thickness.

We can now see why experiments, instead of raising doubts in Aristotle's mind, might easily have confirmed him in his beliefs. Suppose he had taken a glass cylinder six feet long, set it up on end, and filled it in turn with water, wine, olive oil, honey, and molasses. If he had dropped the same pebble down each column of fluid and timed with his pulse how long it took to go $\frac{1}{4}$, $\frac{1}{2}$, and the whole distance, what would he have observed? The answer is: *exactly what he stated.* After an extremely brief initial acceleration, the pebble would fall through the liquid at a steady rate dependent on the viscosity. Double the 'gooeyness', and it would take twice as long: double the force acting—by using a pebble of the same shape and size, but twice the weight—and it would travel the same distance in half the time. In such cases, experience completely supports Aristotle's conclusions: under these conditions, there is nothing seriously wrong in what he says.

What (one can ask) if he had dropped the pebbles through air? Surely he should have seen the limitations of his doctrine? This is doubtful. The time of fall would have been too short to measure with any accuracy: indeed, for all the legends about the Leaning Tower of Pisa, not even Galileo could formulate from direct observation alone the correct law of freely falling bodies; in fact he slowed up the movement by rolling the bodies down inclined planes. Aristotle naturally assumed that the rule of proportion was maintained in air, the short time of fall corresponding to the extreme thinness of air as compared with—say—honey. At the outset, there was no reason to suspect that a body falling through air moved in a significantly different manner.

(iii) *The Impossibility of a Vacuum*

Starting from these principles, Aristotle reached one other conclusion which was to be profoundly influential: namely, that a complete vacuum could not possibly exist in Nature. He argued as follows. If you have any two moving bodies, you can ask how much faster the one is moving than the other. The answer will always be some definite number: 5 × as fast (say) or ½ × as fast. The lengths of time the two bodies take to go from A to B will likewise be in a ratio of 1 : 5, or 2 : 1—again, always some definite number. But suppose one of them is moving in a vacuum, offering no resistance to motion at all; then from the rules of proportion one is led to conclusions which make no sense. For (Aristotle asked) how long would a body take to go from A to B in this vacuum? And the answer appeared to be: No time. The idea of a body moving from one place to another *in no time* appeared to him—reasonably enough—quite inconceivable, and he saw no alternative to denying the original supposition. If the rules of proportion were relied on, there could exist *no* medium offering zero resistance.

To sum up Aristotle's chief conclusions about the motion of terrestrial bodies: (i) Provided that the effort available is large enough in the first place, bodies of different sizes can be shifted in different times by the exertion of given efforts through distances which vary simply in proportion to these things; (ii) all these effects will also vary proportionately, according to the resistance of the medium in which motion is taking place; (iii) but, by these rules of proportion, a body would travel across a vacuum in *no* time—so the possibility of a vacuum has to be ruled out. In all this, motion is treated as a balance between the agent responsible and the resistance to be overcome. So a horse, having got a cart going over level ground, will keep it moving steadily at a rate depending on the roughness of the ground and the lubrication of the axles—at any rate, until the horse begins to get tired. These conclusions are based on careful study of everyday motion, of a kind that we can still observe for ourselves.

(iv) *The Problems of Acceleration and Momentum*

Why did an argument rooted so firmly in common experience lead to conclusions which in the end had to be abandoned? We can see in retrospect that the Achilles' Heel of Aristotle's theory was his treatment of bodies moving against *slight* resistances—of which motion through a vacuum was the extreme case. When a horse sets a cart going or a pebble is dropped through treacle, it reaches its steady or 'terminal' velocity very quickly—so we can give a perfectly adequate description of its movement even though we neglect the initial acceleration. But, as the resistances to motion are reduced, the initial phase of acceleration goes on for longer and longer and becomes—correspondingly—more important. Indeed, until the steady speed is reached, Aristotle's rules of proportion cannot be applied. In the extreme case, i.e. the vacuum, the terminal velocity will *never* be reached, and the acceleration will continue indefinitely. This is in fact the fundamental Newtonian doctrine. Aristotle was quite right to conclude that the *terminal* velocity in a vacuum would be infinite but, because he neglected to analyse acceleration, he put the wrong interpretation on his conclusion.

As we now know, a study of acceleration is of crucial importance, not only in a vacuum, but even in air. Aristotle's arguments, from experience with thick media to conclusions about air, were not reliable. In the case of air, one just cannot neglect the initial acceleration. A man falling from a cliff or an aeroplane will take well over five seconds to reach his terminal velocity (about 120 m.p.h.), and will fall several hundred feet in doing so. This is something Aristotle probably never saw, and certainly never had the opportunity to time! Motions in which the problem of acceleration is important fitted least happily into Aristotle's framework of ideas and resisted for centuries all attempts to incorporate them. Eventually, they were to be the starting-point for a new and more comprehensive system of dynamics. It is no accident that in the period between Aristotle and Galileo, the discussion kept returning to three problems: the motion of a body falling freely through the air, the motion of projectiles (e.g. arrows or cannon-balls) after they have left the projector, and the motion of the planets.

Free fall involved acceleration, but the motion of projectiles and planets raised—in addition—a further question: i.e. what keeps a body in motion once the original motive agent is no longer in contact with it? Aristotle himself recognized the difficulty, but was confident that some explanation could be found which fitted his general scheme. He refers to the projectile problem in his arguments about the vacuum:

> Objects which are *thrown* continue to move, even though the thing which started them off is no longer in contact with them. They keep moving, either—as some say—because of 'mutual replacement' [he means, a turbulence produced when the air displaced from the front of the moving body rushes round to fill up the vacuum left behind], or because the air which they themselves drive forward forces them to move faster than they would if they were simply moving naturally.

To resolve this problem, Aristotle would have had to have a clear conception of momentum: for a flying arrow is a typical case of a body moving 'under its own momentum'. Yet the idea of momentum, like that of acceleration, only became clear gradually in the course of the centuries between Aristotle and Galileo. In his time, it was reasonable to take the horse and cart as the most typical kind of moving body, and to look for active agents keeping projectiles and planets in motion.

The possibility of a terrestrial body moving continuously of its own accord—which was later to be Newton's ideal of natural motion—was in Aristotle's view something wholly *unnatural*. Yet, despite this seeming conflict with modern physics, Aristotle was no fool. As a mathematical ideal, we may be justified in accepting Newton's suggestion—that a body free from all outside interference would move continually of its own accord in a Euclidean straight line for ever. But in real life, as Aristotle insists, that is the last thing that ever happens.

> Nor could one give any reason why, having been set in motion, a body [in a vacuum] should stop anywhere: for why at one point rather than another? The result would be that, either it would remain at rest, or it must continue to move *ad infinitum* until something more powerful got in its way. [Which, he concludes, is absurd.]

If we leave theoretical considerations aside and consider only *practical* possibilities, we cannot take Newton's paradigm seriously.

(v) *The Way Ahead*

Before we turn from Aristotle's dynamics to his astronomy, it is worth asking what obstacles had to be overcome, before a more comprehensive alternative theory could be built up. It was not enough to make more observations or perform more experiments: for, as we have seen, these might just as likely have reinforced as told against Aristotle's views. What mattered, rather, was the way in which the observed facts were interpreted. A more adequate interpretation than Aristotle's called for a complete new battery of concepts and distinctions. Until this was forthcoming, subsidiary hypotheses (such as that of 'mutual replacement') would take care of difficult cases.

What sort of novel concepts were needed? Aristotle had only asked the first and simplest questions about motion. As a result he had drawn too coarse-grained a picture of a very complicated phenomenon. We now realize that one has to distinguish at least half a dozen distinct variables: e.g. initial acceleration, terminal velocity, velocity at any moment, average velocity over a period of time, mass, weight, momentum, and so on. To make all these distinctions, one must define in a consistent mathematical way a whole range of quantities, many of which had not yet been defined—and some of which (it seemed) *never could be defined.* Take one example. In Aristotle's time the very notion of 'velocity' was still liable to get one into trouble. We mentioned earlier his scruples about 'dividing a distance by a time'; and many of his arguments in the *Physics* were designed to avoid logical difficulties to which the idea of velocity could give rise.

Zeno of Elea (for instance) had stated a series of paradoxes apparently demonstrating that the very idea of 'speed-at-an-instant-of-time' involved one in contradictions. If by an 'instant' we mean an infinitely small period of time, then in an instant a flying arrow would not cover *any* distance. Think how it would be if we photographed the arrow in flight: as we made the exposure shorter and shorter, so the picture of the arrow would get less and less blurred—and an entirely 'instantaneous' photograph would freeze the arrow's motion completely. So (it seemed), considered at an instant of time, the arrow could not be said to have a 'speed' at all. And if time is regarded as a succession of infinitesimal instants added together, the paradox results that the so-called moving arrow is at no instant of time genuinely 'moving' at all.

Now this paradox of Zeno's was not just a device for sharpening the wits of philosophy students. At the time he stated it, it represented a real difficulty, which had to be taken seriously by anyone hoping to produce an adequate theory of motion. The concept of velocity—as men understood it in 400 B.C.—led one so easily into inconsistencies that Aristotle was justifiably cautious in his treatment of it. He never clearly saw his way out of Zeno's maze.

Modern physics is no longer troubled by Zeno's paradoxes; but this is only because of the work of seventeenth-century mathematicians. The differential calculus, which they invented, defines 'instantaneous velocity' in a new, sophisticated way, and provides a *consistent* technique for calculating with 'infinitesimal' quantities. One can at last talk about motions occupying 'no time at all' without logical embarrassment. The essential thing (they realized) was not the extreme case, but the manner in which one *approached* the extreme. To pursue our photographic analogy further: in an 'instantaneous' photograph, any flying arrow will appear infinitely sharp, whatever its actual speed. But at all *finite* exposures, however short, the faster arrow will appear more blurred than the slower. So, instead of concentrating on the frozen 'instant', one must look at the way in which the blur is cut down as the exposure is progressively shortened. Speed-at-an-instant can then be defined as a mathematical limit which is approached as the exposure is indefinitely reduced.

THE CELESTIAL MECHANISM

As we saw, Aristotle's programme for physics was a two-stage one. First, one had to establish a general theory of motion, from a study of familiar things close at hand on the Earth; and later, one might hope to apply the physical principles so established to the Heavens. This was the point at which Eudoxos' system of concentric geometrical spheres came in. The geometrical construction Eudoxos employed already hinted strongly at mechanical connections. The task was, to follow up these clues and see where they led.

In the first place, Aristotle insisted, a geometrical scheme could be acceptable only if it satisfied one further condition. It must make *mechanical* sense: i.e. it must tie in with our general ideas about matter and motion. Thinking up a purely geometrical representation of the planetary system was all very well, but for a real understanding more was needed: one had to figure out how the parts of the planetary system were connected together—how it all worked. As an ideal, this was entirely admirable. His ambition was to take Eudoxos' geometrical *description* of the planetary motions ('planetary kinematics') and use it as the basis for a theory about the *interactions* producing these motions ('planetary dynamics'). This was precisely what Newton was to do centuries later with Kepler's more refined planetary kinematics.

For Aristotle's purposes, Eudoxos' account contained one great deficiency. So far as geometry took one, the scheme of twenty-seven concentric spheres could be used to construct planetary orbits very like the ones actually observed. But what *made* the planets move like this, what *forced* them to continue travelling along the intricate tracks, Eudoxos did not say. The most serious gap was between the quartet of four spheres belonging to any one planet and the quartets for the planets on either side. Eudoxos had treated each planet's track as an independent problem: and the resulting scheme was mechanically unintelligible. Aristotle could accept the rotation of the outermost sphere of all—the sphere of the fixed stars, carried round uniformly every twenty-three hours fifty-six minutes. This sphere was the Primum Mobile, which derived its rotation straight from the Divine source of all celestial motion. But how was this rotation (corresponding to Plato's mythical 'Axis of Necessity') conveyed in turn to each of Eudoxos' twenty-six inner spheres? One could not leave the gaps between the different quartets of spheres unfilled.

Aristotle devised, therefore, a mechanism which made consistent mechanical sense of the scheme. Motion must be transmitted, for instance, from the innermost sphere of Jupiter to the outermost sphere of Mars. How was this done? The outermost spheres of *all* the planets moved in the same way—namely, in step with the sphere of the fixed stars—so whatever connection there was between the spheres of Jupiter and those of Mars must cancel out the effects of the three inner spheres of Jupiter; for it is these that together make Jupiter move in its own specific way. The simplest way of cancelling out these effects is to suppose that the mechanical links introduced by the three inner spheres of Jupiter are reversed, one by one, as one proceeds inwards to the outermost sphere of Mars. As Aristotle saw, this would be the case if three additional spheres were interposed between Jupiter and the spheres of Mars, *each* of them moving in exactly the reverse way to *one* of the three inner spheres of Jupiter; similarly for the other inter-planetary gaps.

Suppose all these links in the celestial gear-train rotate uniformly in circles around their own axes and at the same time transmit to the inner spheres a motion whose ultimate origin is the sphere of the fixed stars. You then have a scheme of connections which brings planetary theory into harmony with mechanical sense.

Eudoxos supposed that the motion of the Sun or Moon involves, in either case, three spheres and that the motion of each of the planets involves four spheres. [i.e. 26 in all]

Callippos made the *positions* of the spheres the same as Eudoxos did. But, while he assigned to Jupiter and Saturn the same *number* as Eudoxos, he thought that, if one is to explain the observed facts, two extra spheres should be added for the Sun and two for the Moon; also one extra for each of the remaining planets. [i.e. 35 in all]

But it is necessary, if all the spheres combined are to explain the observed facts, that for each of the planets there should be additional spheres—one fewer than those hitherto assigned—to counteract the former ones and bring back the outermost sphere of the next planet into its proper position; for only thus can all the agencies at work produce the observed motion of the planets. The number of all the spheres—both those which move the planets and those which counteract them—will be fifty-five.

So the fully developed planetary scheme resulting from the work of the classical philosophers in Athens represented the Heavens as a nest of spherical shells, fifty-six in number, with the Earth as its centre. The largest of all was the Divine, self-moving sphere, carrying the fixed stars. Saturn's outermost sphere rotated in step with this sphere, and three more spheres accounted for the planet's own proper motion, which came directly from the innermost of the quartet, to which it was actually attached. Three counteracting spheres linked the smallest of Saturn's spheres with the largest of Jupiter's four; so that Saturn in all had seven linked spheres associated with its motion. Jupiter in turn had seven; Mars, the Sun, Venus, and Mercury had nine each—five to produce their motion, four to 'counteract'; and finally the Moon had five.

Only the outermost sphere had a simple motion: the fifty-five linked spheres carrying the remaining heavenly bodies depended for their motion on the complex linkages which joined them to it. The Earth alone was stationary at the centre of the whole system of concentric shells: the innermost sphere, carrying the Moon, was the boundary between the mortal, 'sublunary' world of the Earth and the unchanging, 'superlunary' world of the Heavens. And this division still echoes on into the poetry of the seventeenth century A. D., whose 'sublunary lovers' bewail the mortality of life and the impermanence of love.

A theoretical scheme of this sort had many attractions. Regarded as a mathematical system, it was geometrically consistent: working from the single fundamental assumption that uniform circular motion was the most appropriate and natural form of motion for unchanging celestial things, it made coherent mathematical sense of celestial motions—which, as watched from the Earth, formed 'highly intricate traceries'. Again, it made mechanical sense, providing an intelligible system of linkages by which the motion of the outermost sphere could be communicated in turn to each of the fifty-five spheres within it.

This was only a beginning. For the same picture of the cosmos also fitted in neatly with Aristotle's ideas about matter and living things. The sublunary world was composed of the four types of matter whose natural motion was limited and vertical—the earthy, aery, fiery, and liquid types—whereas the superlunary world was composed of a fifth distinct type of matter, the so-called 'quintessence' (fifth essence), whose natural motion was circular and unending. The hierarchy of living things, too, stretched from the Divine Heavens, at the outer extremity of the cosmos, down through the lesser, but immortal, intermediate divinities to the world of mortal creatures in the centre. (Aristotle's theories about matter and living things are aspects to which we will return in later volumes.) Nor did Aristotle's scheme do violence to popular religious ideas: earlier natural philosophers such as Anaxagoras had spoken of the celestial bodies in a seemingly derogatory way, but now science could become respectable.

Things in the outermost heaven have no birthplace, nor does the passage of time age them, nor does any kind of change affect the beings whose allotted place is there beyond the outermost motion: free from change and interference, they can enjoy without interruption the best and most independent life for the whole aeon of their existence.

So the divinity of the Heavens, which had been a cardinal point of theology in the Middle Eastern religions, could be retained in Aristotle's system.

Our theory seems to bear out common experience and to be borne out by it. For all men have some conception of the nature of the Gods, and all who believe in the existence of Gods at all, barbarian and Greek alike, agree that the highest place belongs to the deity—presumably because they suppose that immortal things belong together. . . . For in the whole range of time past, so far as our inherited records reach, no change appears to have taken place either in the pattern of constellations in the outermost heaven or in the individual stars making it up. . . . Our present argument demonstrates in addition that the heavenly sphere had no beginning in time and will have no end. Further, it is free from all the inconveniences and constraints of the mortal world; for it does not have to be forced to keep on track, or stopped from moving in some other, more natural way. Such a forced motion would necessarily require the expenditure of effort—the more so, the more unending it was—and would not be consistent with the perfection of the heavens. So we should not accept the old myth that the world needs to be kept secure by some Atlas; . . . it is not only more appropriate to think of its eternity in our way, but also this supposition alone enables us to put forward a theory consistent with popular ideas about the nature of the Gods.

(Notice one feature of Aristotle's cosmology which, at first glance, seems very strange to us in the twentieth century. In spite of all that he has said about the *mechanism* of the heavenly motions, he still insists that the heavenly bodies are *animate*. To understand fully why he says this, we should need to bring in his zoological ideas: his view is roughly as follows. Everything in Nature has a purpose and is animated by a soul fitted to its purpose; Man is the highest of the things on the Earth, just as the sphere of fixed stars is the highest of the things in the Heavens; and the motions of the lower planets bear the same relation to the motions of the outermost sphere as the behaviour of plants and animals bears to that of Man. In the same way as happened with the Pythagorean idea of the 'harmony of the soul', later Christian ideas have made this idea strange, by drawing a sharper distinction between animate and inanimate things than the Greeks found natural. Yet, as the philosopher Whitehead has recently argued, there may be just as much virtue in thinking of the universe as a single giant organism as there is in thinking of it as an enormously complicated machine.)

Still, though Aristotle's cosmology was *respectable*, it was not *dogmatic*. He was prepared, as we shall see shortly, to admit doubts about it—more so than some of his medieval followers—and he was prepared to back up all his assertions with arguments. After all, supposing he had based his system upon *dogmas*, he would have been untrue to his vocation as a philosopher. If the results of his investigation harmonized well with traditional religious principles, so much the better. The words Newton wrote to Bentley about the the theological implications of his own theory might equally have been composed by Aristotle:

> When I wrote my Treatise about our System, I had an Eye upon such Principles as might work with considering Men, for the Belief of a Deity, and nothing can rejoice me more than to find it useful for that Purpose. But if I have done the Public any Service this way, it is due to nothing but Industry and patient Thought.

The first need in each case was to uncover by 'industry and patient thought' the principles governing the operations of Nature, and in doing so to be true, both to reason, and to the evidence of observation.

As a postscript to Aristotle, let us consider one example of his method: namely, the passage in which he argues for the view that the Earth is a sphere. He begins with theoretical considerations: all earthy matter tends to converge into a common centre, so if the Earth came into existence by the congregation of the matter forming it, a spherical form was the result to be expected. (This argument is equally sound on the more modern view, that the planets condensed out of the hot gaseous material from the Sun.)

> The shape of the Earth *must* be a sphere. For all solid [earthy] matter is borne down by its weight until it reaches the centre, and the jostling together of larger and smaller parts would have the result, not that the resulting surface was corrugated, but rather that the different parts would

continue to converge and be pressed together until they had reached the centre. This process can be thought of by supposing that the Earth came into existence in the sort of way described by earlier natural philosophers—except that they treated the downward motion of solid matter as *forced*, whereas in fact the true explanation of this motion is that all heavy things have a natural tendency to move towards the centre. Thus, the matter separated out from the original undifferentiated mixture moved from every side towards the centre in the same way. And it makes no difference whether the parts which came together at the centre were originally distributed uniformly in all directions, or otherwise. If, on the one hand, there had been a similar movement from every direction to the common centre, the resulting mass [of the earth] would obviously have the same shape on every side.... But it does not affect the argument in any way if the amount of matter coming from each direction was not the same. For the greater weight [coming from one direction] finding a lesser one in front of it, must force it onwards, since its impulse is towards the central destination, and its greater weight will drive the lesser one on until this goal is reached.

He then adds arguments based more directly on observation. At all points on the Earth, heavy bodies fall at a precise right angle to a level surface, and not obliquely or along parallel paths: this is intelligible, he rightly argues, only if the Earth is a sphere and all bodies fall towards its centre. Finally, there is evidence from astronomical observations: from the shape of the Earth's shadow, as observed on the Moon's face in a lunar eclipse, and from the way in which stars differ in elevation according to the latitude of the place from which you observe them.

This is further confirmed by the testimony of direct observation. For how else could eclipses of the Moon display segments shaped as we see them? In the ordinary way, the Moon itself displays every month shapes of several different kinds—straight-edged, convex and concave—but in an eclipse the boundary between the light and dark areas is always curved. Since the eclipse results from the interposition of the Earth [between the Sun and the Moon], the shape of this line will correspond to the shape of the Earth's surface, which is therefore round. Again, our observations of stars make it clear, not only that the Earth is spherical, but also that it is a sphere of quite moderate size. For if you travel quite a small way to the North or to the South, the effect on the horizon is easily detected. The stars directly overhead change, and different stars are visible. Some stars can be seen in Egypt and around Cyprus which are not visible further North; and others, which in the North never go out of sight, rise and set in Egypt. All of this goes to show not only that the Earth is spherical, but also that it is of no great size: otherwise so slight a change of position would not have such obvious effect. (For this reason, one should not be too ready to dismiss as incredible the idea that the region beyond the Pillars of Hercules to the West [i.e. the Straits of Gibraltar] are continuous with those beyond India to the East, and that in this way the oceans join up. People quote as further evidence in favour of this view the fact that elephants occur in both these extreme regions—a similarity suggesting continuity.) Further, mathematicians who have calculated the size of the Earth from these observations have arrived at the figure of 400,000 stades. Evidently the matter of the Earth not only forms a sphere, but also one not all that large as compared with the stars.

NOTE: THE SIZE OF THE EARTH'S SPHERE

Aristotle had given excellent reasons for believing that the Earth was spherical in shape: how could the matter again be called in doubt? The knowledge so gained remained a commonplace in those parts of the world—e.g. Baghdad—where the work of Aristotle and Ptolemy was kept alive between A.D. 500 and 1100: if it was lost in Western Europe, that is only one sign showing how dark the European Dark Ages were.

The Pythagoreans had taught that the Earth was a sphere very early in the development of Greek philosophy, and by 400 B.C. this was generally accepted. The mathematical estimates of the Earth's circumference to which Aristotle refers were generally made by comparing the elevation of the same heavenly body as seen from two different latitudes at the same time. The best-recorded calculation is that of Eratosthenes. He chose for his comparison two places of approximately the same longitude, whose distance apart was 5000 stades: Alexandria and Syene, which was to the south of Egypt and on the Tropic of Cancer. On Midsummer Day at Syene, the Sun was directly overhead at noon; but at Alexandria its angle of elevation was one-fiftieth of a circle ($7\frac{1}{5}°$) from the vertical. If the Earth is a sphere whose size is small compared with the distance of

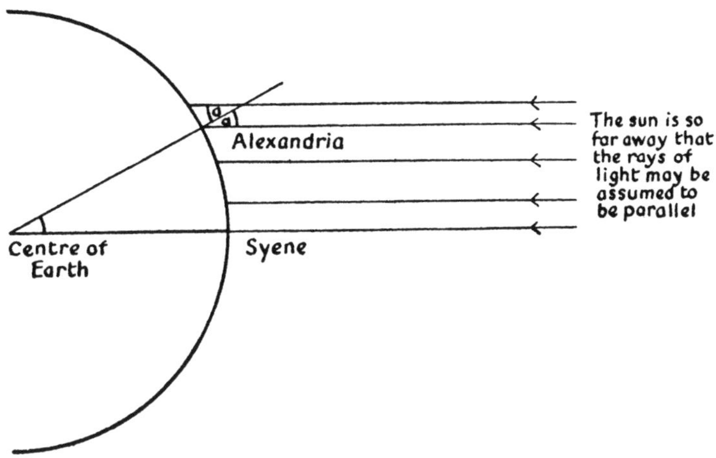

Eratosthenes' estimation of the circumference of the Earth

Angle $a = 7\frac{1}{5}$ degrees—that is, one-fiftieth of a circle of 360 degrees. This angle is equal to the angle subtended at the centre of the Earth by Alexandria and Syene. The known distance between Alexandria and Syene can therefore be considered to be one-fiftieth of the circumference of the Earth

the Sun, then the difference in latitude between Syene and Alexandria can also be taken to be one-fiftieth of a circle. The complete circumference of the Earth will then be fifty times the distance from Syene to Alexandria: i.e. 250,000 stades. Unfortunately, we do not know exactly how many stades there were to a Roman mile: suggested values range from seven and a half to ten. So we cannot compare precisely Eratosthenes' estimate of 250,000 stades with the modern measurement of 24,000 miles. Yet clearly the value obtained cannot have been far from the truth.

Ptolemy later extended the general method. Still taking the sphericity of the Earth as axiomatic, he calculated the distance of the Moon from the Earth. Since, by astronomical standards, the Moon is quite close, its location against the background of the constellations on a given night will depend on the latitude from which it is observed: it will display 'lunar parallax'. Making careful observations from Alexandria, Ptolemy found that it might be displaced in this way as much as 2° from its expected position. On the basis of these observations he calculated that its distance varied around a mean of fifty-nine earth-radii (about 224,000 miles in the modern scale). He subsequently attempted to calculate the distance of the Sun, also, by working out the size of the Earth's shadow when the Moon is eclipsed at a known distance from the Earth. But the method is not easy to apply at all accurately, and the result he obtained (1210 earth-radii) is much too small—being equivalent in modern terms to less than 5,000,000 miles instead of 93,000,000.

FURTHER READING AND REFERENCES

The best general account of the scientific world-picture of the classical Greeks, and particularly of the theories of Aristotle, is to be found in

S. Sambursky: *The Physical World of the Greeks*

For Aristotle's ideas about dynamics, one may consult

T. L. Heath: *Mathematics in Aristotle*
F. M. Cornford: *The Laws of Motion in Ancient Thought*

Useful general discussions will be found also in

M. Clagett: *Greek Science in Antiquity*
T. S. Kuhn: *The Copernican Revolution*

in addition to the books by Burnet and Dreyer referred to earlier.

4

Doubters and Heretics

ARISTOTLE'S theories were never sacrosanct in the ancient world—least of all to Aristotle himself. It would have been very surprising if they were. In a period of some 250 years, the front line of Greek natural philosophy had been pushed forward again and again. The small beach-heads established by Thales and Pythagoras had been enlarged, united, and expanded through the work of their successors, to form a secure base for future operations. So the Greek philosophers were entitled to be confident about the general direction of their work; but equally, where so much had been changing so quickly, they could hardly suppose that the last word had been said. Aristotle's synthesis was, no doubt, extremely attractive. But for philosophers convinced that reason could uncover the principles of Nature, its merits must depend on the arguments and evidence supporting it. Centuries later, Aristotle's conclusions were to be integrated into the wider framework of Christian theology—a dogmatic system whose main purposes were quite other than his own. For the moment, they were at the mercy of fresh evidence and better arguments.

Scarcely an item in the whole system escaped criticism from one or other of Aristotle's successors. Plato himself had already questioned the immutability of the Heavens:

> The genuine astronomer will admit that the sky and all it contains have been framed by their maker as perfectly as such [material] things can be. But . . . he will not imagine that these visible, material changes go on for ever without the slightest alteration or irregularity.

Aristotle, for his part, eventually had doubts about the theory of concentric planetary spheres. His difficulties were shared by later astronomers, and during the next four centuries planetary kinematics became far more complex than it had been for Eudoxos. Two generations after Aristotle, Aristarchos of Samos even revived the Pythagorean suggestion that the Earth—instead of being immobile—travelled in a circular orbit: in his view the Sun and the fixed stars alone were stationary. In dynamics, again, the problem of fitting freely-falling bodies and projectiles into Aristotle's theory was never far from people's minds; and, as time went on, these cases became more difficult rather than less. Yet no one came forward with a coherent and consistent theory anything like as comprehensive as the one Aristotle had suggested.

In the face of all these difficulties, why did the Greek philosophers not throw Aristotle's theory aside at once? Why were they not content to suspend judgement on the questions he raised? Over this question, it is easy to be too puritanical. There is a time to suspend judgement; and there is also a time to reach a compromise. The difficulties which beset Aristotle's theory were genuine enough, but we must also recognize that it had its strong points. For the different parts of the scheme interlocked very beautifully, meeting simultaneously many of the demands which men require a cosmological system to satisfy. Unless we acknowledge its merits, we shall never understand why the theory remained for so long the cosmological foundation of European thought.

There had to be amendments, certainly; and some of these in due course ran counter to the basic principles on which the theory had been first constructed. This kind of intellectual opportunism can be found in the scientific thought of all periods. Men are often obliged to introduce crucial modifications into a theory, without wholly abandoning it, confident that in the end some way will be found to reconcile them with the basic principles of the theory. Indeed, unless someone comes along with an equally comprehensive alternative system, that is all they can do. No one likes to be out on such an intellectual limb, yet there are times when we have to reconcile ourselves to this position. For the moment, Aristotle's theories of motion and the cosmos represented the best general framework of ideas available, and his successors legitimately hoped—by suitable tinkering—to make them apply in all cases.

The first important addition to Aristotle's theory of motion foreshadowed the modern idea of momentum. The earliest surviving statement of the need for some such idea dates from the sixth century A.D.: it appears in the commentary on Aristotle's *Physics*, written by Ioannes Philoponos, the Byzantine. Still, there is reason to think that the substance of Philoponos' theory goes back to Hipparchos in the second century B.C.: it has all the character of the most mature and original scientific thought of the Greeks.

Philoponos discusses both falling bodies and projectiles. He points out clearly the limited scope of Aristotle's 'rules of proportion'. Whatever may be the case in viscous media, such as honey, it is a mistake to suppose that bodies fall through *air* at speeds proportional to their masses. It would be a fair conclusion from Aristotle's principles, he said, that

> when bodies of different weight were moving through the same medium, the times required for their motions would be inversely proportional to their weight. For example, if the weight were doubled, the time needed would be half. But this is completely erroneous, as may be confirmed by actual observation more effectively than by any sort of verbal argument. For if you let two weights fall from the same height, one of them being many times as heavy as the other, you will *see* that the times required for the motion are not in a proportion dependent on the ratio of their weights, but that the actual difference in time is very small.

Philoponos is here imagining, a thousand years in advance, the experiment usually credited to Galileo. The evidence that Galileo used the Leaning Tower of Pisa for dynamical demonstrations is actually very weak: in any case, his most mature theories date from long after he had left Pisa for Padua and Florence. If anyone, it was probably Simon Stevin of Bruges who introduced Philoponos' argument and experiment to Europe. What is important, of course, is this: that Philoponos' demonstration does *not* by itself refute Aristotle. It merely adds one more case to those already mentioned by Aristotle in which the rules of proportion do not apply: e.g. the fact that a ship which 15 men can shift 15 feet in an hour may be too big for one man to shift one foot in the same time—or to shift at all.

Similarly for projectiles: here again, Philoponos argues that common experience reveals an inadequacy in Aristotle's theory. In the case of a stone or arrow thrown through the air, Aristotle had seen no alternative but to suppose that some *external* cause—such as the disturbance in the air produced by the flying body—was responsible for its continuing in motion. Philoponos throws out the suggestion that, in this case, the cause of the continued motion can more fruitfully be thought of as *internal*. Again he appeals to an imaginary experiment. If air-disturbance had sufficient power to keep a body moving at high speed, then by blowing air at a stationary body you should be able to get it started in the first place; but obviously you cannot.

> When one projects a stone forcibly, does one compel it to move contrary to its natural direction of motion by disturbing the air behind it? Or does the thrower also impart some [internal] motive power to the stone? If he does not impart any such power to the stone, but moves it merely by pushing the air, and if the bowstring moves the arrow in the same way, what advantage is it for the stone to be in contact with the hand, or the bowstring to be in contact with the notched end of the arrow?
>
> In that case, one might place the arrow or stone on the top of a stick, only touching it along a thin line [to cut down friction]; and then, without there being any direct contact between the projector and the projectile, one could use machines to set a large quantity of air in motion behind the body. Clearly, the more air moved and the more forcibly it was moved, the more it should in its turn push against the arrow or stone and the further it should hurl them.
>
> The fact is, that even if you placed the arrow or stone on a support quite devoid of thickness and set in motion with all possible force all the air behind it, the projectile would still not move as much as a single cubit. . . . From these and from many other considerations, we can see how impossible it is for this to be the true explanation of the continued motion of projectiles. Instead, it is essential to suppose that *some incorporeal power* is transferred from the projector to the projectile, and that the air set in motion contributes either nothing at all or else very little to the projectile's movement.

This 'incorporeal power' is the first recorded forerunner of our own term 'momentum'.

Yet, even though Philoponos put his finger very precisely on the points at which Aristotle's dynamics faced the greatest difficulties, he was in no position to reject the theory entirely, or to reconstruct it from the ground up. Mechanics did not stand still after Aristotle—Archimedes did important work on hydrostatics and the principle of the lever, and Strato carried Aristotle's analysis of the rules of proportion a good deal further. Nevertheless his theory remained the starting-point for later work; and a radical reconstruction was not forthcoming until the seventeenth century A.D.

AMENDING THE ASTRONOMY

Aristotle himself seems to have begun the patchwork on the astronomical side, though only second-hand accounts now exist of the difficulties he tried to deal with. The question that worried him was: 'Are the heavenly spheres really concentric?' And his doubts sprang from certain apparent variations in the sizes of the heavenly bodies—notably: Venus, Mars, and the Moon. It was tempting at first to put these down to variations in the atmospheric conditions; but this possibility was soon ruled out. The only other natural explanation was that the distances of these bodies from the Earth were markedly variable. Yet such an explanation was quite inconsistent with Eudoxos' picture. If the spheres carrying the Sun, Moon, and planets were all exactly centred on the Earth, the distances from a terrestrial observer to all those bodies would be effectively constant. (Even for the Moon, the nearest body, the extreme variation in distance would be only about 1 per cent.)

The seriousness of this problem is clear from the following passage. Simplicius (sixth century A.D.) is quoting Sosigenes (second century A.D.):

> At some times the planets appear to be quite near us; and at others they appear to have receded. In some cases this variation is easily visible: thus, in the middle of their retrogradation, the planets Venus and Mars seem to be many times as large as usual—so much so that, on moonless nights, the light from Venus actually makes bodies cast shadows. It is clear also, even to the unaided eye, that the Moon is not always at the same distance from us, because even when the atmospheric conditions are the same it does not always appear the same size.

This is confirmed if we study the Moon with an instrument. The disc which, when placed at a fixed distance from the observer, exactly obliterates the Moon will have a breadth at one time of 11 fingers, and at another time of 12 fingers. There is further evidence for this view [i.e. the varying distances of the heavenly bodies] in the records of total eclipses of the Sun. When the centre of the Sun and the centre of the Moon happen to be in a straight line from one's eye, what one *sees* is not always the same. On one occasion, the cone subtended by the Moon, and having its apex at one's eye, entirely includes the Sun, so that the Sun remains invisible for a marked period of time; yet on another occasion, so far from this being the case, a definite ring of light may remain visible all round [the Moon] even in the very middle of the eclipse. [This is known as an 'annular' eclipse.] From this we are forced to the conclusion that the apparent variations in size of the heavenly bodies, when observed under the same atmospheric conditions, are a consequence of their varying distance from us.

Aristotle showed that he was aware of this when in the *Physical Problems* [a lost book] he discussed astronomers' objections, based on the fact that the sizes of the planets do not always appear the same, to the hypothesis of the concentric spheres. In this respect, he himself was not entirely satisfied with the idea of the rotating spheres, although the supposition that they moved uniformly about the centre of the cosmos and were concentric with it was one he found attractive.

Other difficulties arose in due course, when the planetary records were scrutinized more closely, and checked against the implications of Eudoxos' system. As well as these variations in distance, there were certain variations in speed which were also not accounted for by the theory of concentric spheres. Some alternative theories were, indeed, already being put forward in Aristotle's own lifetime. Herakleides of Pontos, who was four years older than Aristotle, made two important suggestions. (i) The first related to the curious fact, remarked on by Plato, that the Sun, Mercury, and Venus move always in conjunction; the two small planets sometimes appearing in front of the Sun, sometimes behind it. Herakleides accounted for this by supposing that these two planets accompanied the Sun like 'Moons' in its annual journey round the Earth. (ii) The daily motion of the stars, he argued, could just as well be due to the Earth's rotating on its own axis, as to a rotation of the sphere of fixed stars: everything would *look* the same, and a much less violent movement would be required.

Once these two suggestions had been put into circulation, two others naturally followed. The less drastic was that of Apollonios. Herakleides' first thesis implied that Mercury and Venus travel round the Earth, not in simple circles, but in orbits which are a combination of two separate circular movements—all three bodies share in the larger circular motion of the Sun, but in addition Mercury and Venus travel round the moving Sun in smaller 'epicycles' (or circles-on-circles). Seen from the Earth, the resulting track will appear to form a loop, whenever Mercury or Venus stops

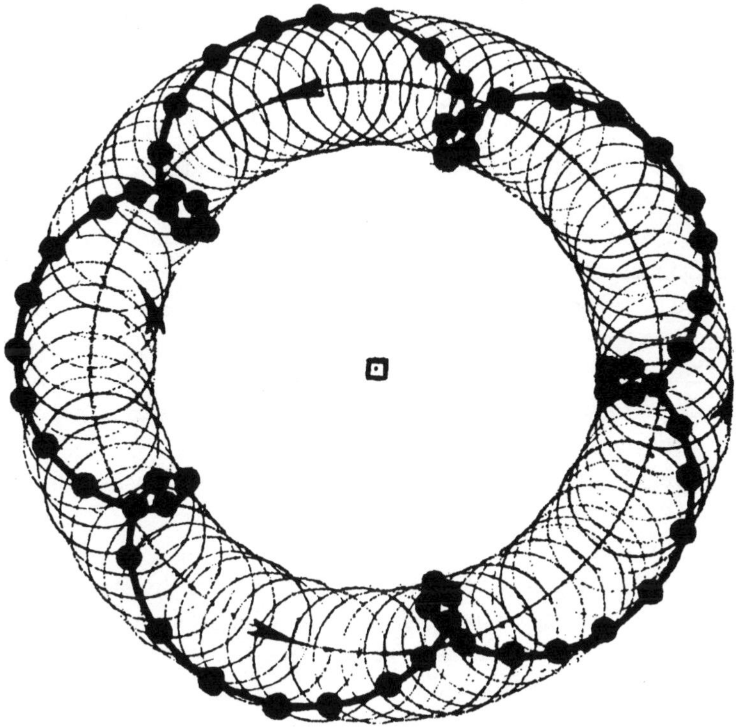

The retrograde motion of the planets explained by the combination of two circular movements

The planet travels along a main circular path—the deferent—and, superimposed on this, at the same time moves in the epicycle and retrogrades across the Sun. Apollonios saw that a similar idea would account for the retrogradations of the outer planets. Suppose that Mars has an orbit, which is likewise a combination of two separate circles: a large 'deferent' circle whose centre is the Earth, combined with a smaller 'epicycle' whose centre is, not the Sun, but a steadily moving point on the deferent circle. If one chooses appropriate radii and speeds for these two rotations, the sum of these two motions will be a track looped in the way we actually observe. Furthermore, the result will be that Mars, in the middle of its retrogradation, will be much closer to the Earth than at other times: so the problem of the apparent change in its size, as described by Sosigenes, is solved at the same time.

ARISTARCHOS' HELIOCENTRIC THEORY

However, Aristarchos went even further. Herakleides had made the stars stand still: according to him, their daily motion was an optical illusion produced by the Earth's *rotation* on its own axis. Aristarchos extended this to the annual motion of the Sun—which (he said) was another illusion, produced by the Earth's *translation* along its own circular orbit. The Sun, not the Earth, was at rest in the centre of the starry sphere. This was basically the same view that Copernicus, 1700 years later, advocated in greater detail and with fuller arguments.

Aristarchos' suggestion was not taken very seriously. Although one other of his astronomical treatises has been preserved—that has to do with the sizes and distances of the Sun and Moon—we have only brief, second-hand accounts of his heliocentric hypothesis. A generation later, Archimedes knew of it and referred to it in his book *The Sand-Reckoner*, which deals with the magnitude of the universe:

> As you know, the name 'cosmos' is given by most astronomers to the sphere whose centre is the Earth, and whose radius is equal to the distance between the centres of the Earth and the Sun: this you have seen in the treatises written by astronomers. But Aristarchos of Samos published a book of speculations, in which the initial assumptions led to the conclusion that the whole universe is very much larger than what is now called the

cosmos'. He supposes that the fixed stars and the Sun are stationary, that the Earth travels round the Sun along the circumference of a circle . . . and that the sphere of the fixed stars is so vast in extent that—by comparison—the supposed circular orbit of the Earth is, in effect, no larger than the central point of a sphere compared with its surface.

Aristarchos evidently tried hard to make the heliocentric view acceptable to his contemporaries; but without success.

Why did he fail? People sometimes point to this episode as evidence that the Greeks took too much notice of 'philosophers' and not enough of 'scientists'—as though there were something intrinsically scientific and obviously true about a heliocentric view, which should have made it immediately preferable to the doctrines of Eudoxos and Aristotle. This is a plain misreading of the situation. Quite apart from the question whether Aristarchos was more or less of a 'scientist' than Aristotle, the weight of the scientific evidence of this time was against the heliocentric view. The suggestion was certainly worth putting forward, but it would have been unscientific of the Greeks to have accepted it as a well-founded theory. It was, in fact, a speculative hunch; and Aristarchos' contemporaries treated it as such. They mentioned it as an interesting idea, pointed out the leading objections, and passed on to the next question.

The objections to the heliocentric theory were of three kinds. The most important arguments were scientific ones, some of which came from physics and others from astronomy; but, for good measure, the general religious sentiment also counted against the heliocentric view.

The astronomical objections were concerned with questions about the fixed stars. (So far as the planetary system went, existing observations were equally consistent with a heliocentric or a geocentric theory.) The most striking single fact about the visible Heavens is this: that, from all parts of the Earth and all times of year, the relative sizes of the different stars remain always the same, while their angular separations and the patterns they consequently form in the constellations never visibly alter in the slightest. From this fact it was natural to infer three things: (i) that the stars were all the same distance from the Earth, (ii) that their distance did not alter—i.e. that the Earth was at rest compared with the stars—and

(iii) that this distance was enormously larger than the size of the Earth. For, if the Earth's distance from the stars were variable, one would expect visible changes in the patterns of the constellations as a result. The stars would then be like distant objects seen from a train, whose pattern changes as the train moves along. (This astronomers call 'stellar parallax'.) And, if the Earth's size were an appreciable fraction of the cosmos, such changes would result, even as one travelled around on the Earth's surface. Yet, in fact, no stellar parallax of any kind was detected, even with the aid of telescopes, until the nineteenth century A.D. (Plate 6.) Furthermore, added Ptolemy,

> There is clear evidence for this in the fact that those planes passing through our eyes which we call 'horizons' always cut the celestial sphere exactly in two. This could not happen, if the size of the Earth were appreciable compared with the distance of the Heavens; since, then, the only plane capable of cutting the celestial sphere into equal halves would be the one passing through the centre of the Earth. Whereas, horizontal planes drawn through any point on the *surface* of the Earth would cut off a smaller part of the heavenly sphere above and a larger part below.

On all counts, therefore, the Greek astronomers were prepared to allow that the sphere of the fixed stars was very large.

Yet Aristarchos' doctrine required the universe to be, not just large, but incredibly large. For, to account on his theory for the unchanging patterns of the constellations, one had to suppose that the whole diameter of the Earth's annual orbit was a mere point compared with the distance of the stars: the stellar patterns were unchanged as one travelled, not just from Gibraltar to India, but across the whole six-months' width of the Earth's yearly track. Though the Greek scientists were ready enough to consider original and surprising speculations, this went too far for most of them. Aristarchos admitted (as Archimedes tells us) that on his theory the whole orbit of the Earth was no larger than a point in comparison with the sphere of the stars, and the Greeks found this implication frankly incredible. The general public today is so innured to astronomical statistics that people will say unthinkingly, 'The stars are an incredible distance away'—and leave matters at that. Greek astronomers were more particular about their arguments, and declined to believe the incredible. In their view, the price of Aristarchos' theory was not worth paying.

This was not just lazy-mindedness, either. If they had pursued the argument further, they would not quickly have found reason to change their minds. For, given a heliocentric theory, difficulties next rise over the apparent size of the stars. As an example: if the stars are really so far away, how is it that we can see them at all, and why have they not shrunk to invisibly small points? Or are they all of fabulous size? These difficulties came to the fore again in the sixteenth century A.D., after Copernicus had revived Aristarchos' theory: Tycho Brahe found them insuperable, and it was not until the time of Galileo that a way was found around them.

Astronomical difficulties alone would surely have been enough to sink Aristarchos' theory for the time being. No one disputed the attractions of the theory, which allowed one to suppose that both the sphere of the fixed stars and the Sun were at rest: the only question was, what intellectual price was worth paying for these merits. But the astronomical arguments were, in any case, bolstered up by dynamical ones.

Few people, even today, find it easy to come to terms with the implications of a belief in the Earth's daily and annual movements. And indeed, one's mind at first boggles at the thought of the whole material globe, together with us men, our houses, and possessions, not to mention ships, animals, and trees, hurtling every twenty-four hours over a distance of some 2,000,000 miles, and all the while spinning round eastwards at up to 1000 miles per hour. Need we wonder that, for reasons of dynamics, Ptolemy found both the rotation and the translation of the Earth quite incredible? To begin with, he asked, why did its translation not leave everything else behind?

It is normally supposed that the whole body of the Earth, being so much larger than the things which it carries and which fall on to it, can absorb the shock of their fall without itself acquiring any motion at all. But if the Earth shared a common motion along with all other massive bodies, it would soon leave them behind—travelling faster on account of its greater size—so that animals and other heavy bodies would be left with no more visible means of support than the atmosphere; and before long the Earth would even disappear out of sight. These consequences are too ridiculous even to imagine.

As for the rotation of the Earth: how could any terrestrial body —particularly something like a cloud, up in the atmosphere—ever move eastwards over the surface of the Earth, if the whole thing was already rushing in that direction at getting on for 1000 m.p.h.? This, Ptolemy argued, would mean admitting

that the Earth has a movement, on account of its revolution, more rapid than any of those movements which take place around it, seeing that it has so far to rotate in so short a time. In that case, all those bodies not actually carried along by it must appear to have a movement contrary to the Earth's own movement. Neither clouds, nor projectiles, nor flying animals, would ever appear to be moving Eastwards; since the Earth would always travel faster in that direction than they did, and would outstrip them by its own Eastward movement. The result would be that all other bodies would seem to be falling back towards the West.

The dynamical consequences of Aristarchos' view were unacceptable without a revolution in dynamical ideas.

Compared with these scientific objections to the heliocentric view, the religious difficulties seem scarcely worth mentioning. We know only that Cleanthes accused Aristarchos of impiety 'for setting the Hearth of the Universe in motion', and there is no suggestion that anything came of the accusation—such as the actual imprisonment Anaxagoras had earlier suffered. All three kinds of objection remain forceful right up to the time of Newton. Even Copernicus had difficulty in meeting them and between 1450 and 1687—when Newton's dynamics finally resolved the conflict—the number of convinced Copernicans was never large.

So we really need not be surprised that the Greeks remained sceptical about Aristarchos' suggestion: rather, we should congratulate them on their good sense. In judging them as scientists—as rational interpreters of Nature, that is—the important thing, surely, is not to ask how many conclusions they reached which we still accept, but rather how far their conclusions were supported by the evidence then available. In so far as they allowed their judgement to be influenced by the weight of the evidence, they can be said to have thought scientifically. In the light of subsequent developments, we can dispose of their arguments for the geocentric theory. But, as matters then stood, they had equal scientific justification for rejecting Aristarchos' heliocentric speculations.

5

Physics Loses Momentum

UP TO this point, the scope of science had been continually widening. The Greek faith in the rationality of Nature, originally born in Ionia, seemed to have taken firm hold—at any rate among the philosophers. On the face of it, there was no reason why Greek science should not expand to embrace all aspects of the natural world and all sorts of problems. But over the next 500 years things took a new, and in some ways a paradoxical, turn. After 250 B.C. the intellectual centre of the Greek world shifted to Alexandria, and there the opportunities for scientific work were in many ways more favourable than they had been earlier in Athens. Yet science gradually lost its momentum, and scientists lost faith in their methods. New sorts of questions were rarely asked: urgent but intractable problems were shelved. By the time of Ptolemy, cosmology, physics and mathematical astronomy, brought together by Aristotle, had fallen apart again; and the science of the sky had become once more only a collection of mathematical techniques.

More than 1000 years later the situation was much the same. Copernicus was still trying to improve on Ptolemy's mathematical techniques, though he hoped to remove the worst irrationalities from his methods: in the course of his work, he was to revive Aristarchos' suggestion of a moving Earth. Meanwhile the framework of Aristotle's astronomy had been transformed into a fundamentally religious picture of the cosmos. The basic physical questions, about the mechanical workings of the planetary system—which Aristotle had raised and Ptolemy had ignored—were not investigated again seriously until the seventeenth century. As we shall see, mathematical astronomy after Copernicus was back again at the level of achievement to which Ptolemy had raised it: the *physics* of the planets—the theory of the laws and causes governing their motions—was, on the other hand, only back to the point where Eudoxos had left it.

FOUR QUESTIONS

We are faced at this stage by four intriguing questions. Two of them are general ones: they concern the development of science as an intellectual discipline, and in particular the factors affecting the pace of this development. These questions we shall only touch on here, since we shall be discussing them more fully elsewhere. The other two questions are more specifically concerned with astronomy, and these will be our main topic. These four questions are as follows:

(i) What factors, social and intellectual, were responsible for the Greek scientific tradition losing its original character, and eventually being cut off entirely?

(ii) What original contributions did Ptolemy make to astronomy?

(iii) Why was it that, during the thirteen centuries between Ptolemy and Copernicus, the science of astronomy did little more than mark time, so that Copernicus' first question takes up immediately the issues raised by Ptolemy's last?

(iv) What exactly did Copernicus criticize in earlier astronomical ideas, and how revolutionary were his own innovations?

In the present chapter and that which follows, we shall discuss each of these four questions in turn.

Clearly, the first and third of these questions have to do, not only with the internal structure and development of science, but also with those external influences which affect its development in one way or another. The ways in which these external factors alter its speed and direction of development can sometimes be analysed in part—at any rate in retrospect—but are not the sort of thing one can seriously attempt to predict beforehand. Political and social changes, economic factors, religion, chance, intellectual faith, and intellectual honesty: all these are important. So long as the manner in which men's ideas about Nature evolve is not fully understood, we shall not be able wholly to explain these influences: we can only demonstrate that they had their effect. In this present book, therefore, we shall deal with such general questions only for the light they throw on the historical development of astronomical and dynamical ideas. Still, the following points can at this stage usefully be borne in mind.

First: when one considers what happened to intellectual life in the Greek world between 300 B.C. and A.D. 200, one very striking thing is the way in which the scholars become dispersed over the whole region. During the fifth and fourth centuries B.C., it was in the schools and academies of Athens that the central problems of Nature had been argued and re-argued. During the next century and a half, lines of research opened up at Athens were carried further by men in Sicily, Rhodes, and Alexandria. But from then on for 600 years the story is one of individual scholars arising, who are capable of mastering the tradition, and making some original contributions to it. Co-operative attacks on the outstanding problems of science become rare: it is the individual that counts.

Secondly: for many reasons, political and social, personal contact between different scientists seems to have become less frequent. Now, in the twentieth century, the scientific tradition might continue even if all our contemporary scientists never spoke to one another again. We have the printed page. Science would certainly be devitalized, since its health depends on its ideas being argued out in open debate; but it might yet remain alive, and develop. On the other hand, when the inheritance of the intellectual tradition and transmission of ideas both depended on personal contact, between master and master, and between master and pupil, any interruption in the chain of personal contacts was liable to be final.

Thirdly, and probably most important: during these centuries, scientists gradually lost faith in themselves and in their methods of thought. It was not only that the first enthusiasm had gone: in addition, from 100 B.C. on, men began to doubt more and more whether after all rational inquiry alone *could* uncover the workings of the Heavens. And since this problem had been something of a touchstone for natural philosophy, failure in this direction had wider repercussions. By A.D. 200, astrology had recovered all the ground it ever lost, and had effectively displaced rational astrophysics. Greek astronomers began to limit their ambitions, and to concentrate on doing those things that they were already good at doing. They became satisfied with making small amendments to existing mathematical theories; filling in details rather than branching out in new directions. Since the progress of science demands that we should always be trying to solve the problems that have so far defeated us, and not just go on applying the techniques we already have, later Greek scientists in this way contributed positively to the decline of their subject.

THE POLITICAL BACKGROUND TO LATE GREEK ASTRONOMY

The political upheaval which transformed the Greek world began in Aristotle's own lifetime. Despite intermittent pressure from Persia in the East, the city-states of the Greek mainland had maintained their independence until about 350 B.C., fighting each other rather than the 'Barbarians'. Towards the end of the fourth century B.C., however, the greater part of Greece was forcibly unified, being absorbed by the expanding kingdom of Macedonia, in the North.

Alexander, its most famous ruler, was a man of wide interests. When in 334 B.C. he started on his crusade against the Persian Empire, he set off with the grand ideal of creating a world-union of states, in which all men would be regarded as equal, Greek and Barbarian alike. The technical staff accompanying him—private secretaries, historians, communiqué-writers, surveyors, botanists, geographers, and other scientists—was probably unequalled until Napoleon went to Egypt in the nineteenth century A.D. Some of these men seem to have been attached to the expedition at Aristotle's own suggestion: he never neglected an opportunity of getting reports on animals and plants of distant lands. His nephew, Callisthenes, was one of the staff historians, also charged with the duty of sending back to Macedonia any information he felt would be of interest to Aristotle.

For science, the most significant result of Alexander's career was his foundation of the city which was soon to displace Athens as the centre of gravity of the intellectual world. His first aim had been completely fulfilled. He destroyed the political power of Persia utterly, and made Egypt and Mesopotamia provinces in a Greek-ruled federation; indeed, before his followers were finally overcome by homesickness, he had marched as far as India and Central Asia. Yet his dream of creating a prototype 'United Nations' never came true. After his premature death, his generals divided his empire into four parts, each of which became a separate kingdom. For our story, the most important of these regions is Egypt.

When in Egypt, Alexander founded the new port and city of Alexandria on the delta of the Nile. The first Macedonian king of Egypt, Ptolemy Soter, ruled from 305 to 283 B.C. and established the dynasty which was to last until Cleopatra's death. He set up in Alexandria the famous 'Museum'. Like many modern colleges, this was nominally a religious foundation, dedicated to the Muses, but it soon developed into a kind of state-supported university and research institute. Associated with the Museum was the equally famous Library, which very soon built up a collection of half a million volumes, as a result (it was said) of buying up the personal book-collections of famous scientists such as Aristotle. These two institutions, the Museum and the Library, together formed the best-equipped centre for advanced study ever to exist in the Greek world, and from about 250 B.C. specialized work of a scientific kind shifted more and more from the mainland of Greece to Alexandria. It is true that some of the most talented individual scientists lived and worked elsewhere: Archimedes (287–212 B.C.) at Syracuse in Sicily, and Hipparchos (180–125 B.C.) on the island of Rhodes. Yet the facilities provided in Alexandria had an understandable attraction for scholars—the attraction that, for similar reasons, the United States has exerted in the twentieth century on scholars in the rest of the world.

Everyone who has read Shakespeare's play *Antony and Cleopatra* knows of the eventual Roman conquest of Egypt, in the last century B.C. Yet this political change had little effect on the academic life of Alexandria. Almost 200 years later the great astronomer Claudius Ptolemy worked at the Museum, and it was still a centre of mathematical research even in the fourth century A.D. From then on, however, conditions for scientific work rapidly deteriorated. Early in the fourth century, Constantine, the first Christian Emperor, had established his own new city of Constantinople beside the Bosphoros, on the site of the Greek town of Byzantium—the city which we know nowadays as Istanbul. By A.D. 500 this had become the capital of the Eastern Roman Empire and was isolated from Rome by the German invasion of Italy. It soon turned into a centre of religious orthodoxy, becoming more and more intolerant as political pressures became more severe. With orthodoxy imposed from above, and fanaticism at work all around, the scholars of Alexandria began to look away from the Greek world, towards Persia and the East, in the hope of finding a more favourable atmosphere for their work.

THE SCIENTIFIC BACKGROUND: THE RETREAT FROM PHYSICS

In the years immediately after Aristotle, Greek science did not substantially change its direction. For a few years after his death astronomical and mechanical problems seem to have been neglected: his immediate successor at the Lyceum was a botanist. But in 287 B.C. Strato took over the direction in turn, and interest in problems of movement revived: to such an extent that the weak points in Aristotle's account were fully recognized. Strato was quite clear, for instance, that bodies left to fall freely to their own natural level do not move with uniform velocity but accelerate, and as evidence of this he cited the following familiar observation:

> If one watches water falling down off a roof through a considerable distance, the flow at the top will be seen to be continuous, but at the bottom of the stream the water falls in separate splashes. This could never happen unless the water was falling more swiftly through each successive distance than it had through the earlier ones.

About this time someone connected with the Lyceum wrote the treatise known to the Middle Ages as 'Aristotle's *Mechanics*'. This was the first systematic attempt to translate Aristotle's common-sense generalizations into mathematical equations. The problems created by doing this were in due course to be one of the main starting-points for the mathematicians of Mediaeval Europe.

Meanwhile, in mathematics, Archimedes was introducing quite novel methods of argument, whose full possibilities were not recognized until the sixteenth century A.D. Some of these arguments were highly ingenious: for example, those in which he applied his novel idea of the 'centre of gravity' of a body to solve problems in statics—treating the weight of the body, for purposes of calculation, as concentrated at a point rather than spread throughout it. This intellectual step, of replacing an extended weight by a localized one, has been effective again and again in the history of scientific thought. Newton's own work on gravity was held up until he was satisfied that one could treat the gravitational action of a large body like the Earth in the same way. Archimedes' other main idea in mathematics was no less significant. We have seen that dynamics was crippled for lack of any adequate idea of 'velocity-at-an-instant'; and this could be defined consistently only as a mathematical 'limit'. Now the possibility of arguing in terms of 'limits' was first clearly demonstrated by Archimedes, and this demonstration was a major step towards the development of the infinitesimal calculus. (At the end of this chapter, as an example of his method, we shall explain the Archimedean proof of the formula for the area of a circle.)

We are now, however, entering the period of the individual. Sometime around 150 B.C., Hipparchos of Rhodes was still writing on the crucial problems of dynamics, for instance, in his treatise *On Bodies Falling Under Their Own Weight*. But he left no successors and we have only second-hand accounts of his work. From these it appears that he anticipated the later views of Philoponos and was working towards the idea of momentum. Once again, the form in which he posed the problem of free fall involved questions very similar to those which were taken up again by mathematicians in the twelfth and thirteenth centuries A.D.

Had Hipparchos and Archimedes left behind them pupils equally concerned with the intractable problems of dynamics, the difficulties which held everything else up until the seventeenth century might have been overcome much sooner. As things were, astronomy rapidly outran physics. Men's general ideas about matter and motion were still too crude for them to get a physical understanding of the celestial motions in terms of laws and forces. From then on, the situation in this respect got worse, not better. In observational and predictive astronomy, the period up to A.D. 150 was one of striking progress. In astrophysics, on the other hand, they were years of gradual retreat.

The ultimate ambition of natural philosophers was still to give a physical explanation of the celestial motions; but this ambition became less and less of a genuine hope. We find it stated once more as an ideal, by Geminos, in the first century B.C.:

> In many cases the astronomer and the physicist set out to prove the same conclusions: e.g. that the sun is very large or that the earth is spherical in shape—but they will not go by the same route. The physicist will explain everything by appeal to the natures and substances of things, to the forces involved, to the utility of the existing state of affairs, or to the theories of change and development: whereas the astronomer will account for the facts geometrically or arithmetically, or by relating the amount of movement found by observation to the time needed for it to take place. Again, though the physicist will often find the explanation by identifying the force responsible, the astronomer—looking only to the external conditions of the phenomenon—is in no position to judge of the underlying cause, e.g. of the spherical shape of the earth. Sometimes, indeed, he has no particular interest in finding out the cause, e.g. of an eclipse; and at other times he confines himself to stating hypotheses or suppositions by which the observed phenomena can be fitted together mathematically.
>
> How, for instance, is one to explain the fact that the sun, moon and planets give the appearance of moving in an irregular way? To this we may answer [as astronomers] that by assuming that the circular orbits of the celestial bodies are eccentric, or that they describe epicycles, the seeming irregularities in their motions will be accounted for. But this is not enough: we must also look into the question, in how many different such ways the observed phenomena could be brought about; so that we may bring our [mathematical] theory of the planets into line with an explanation of the underlying physical causes which is theoretically admissible.

So, in the time of Geminos, the unification of physics with mathematical astronomy still seemed a reasonable goal. Two hundred years later the problem appeared quite desperate; and in fact this goal was not to be reached until the time of Newton. The contrasting attitudes of Hipparchos, who was still hopeful, and Ptolemy, who had written astrophysics off, show how drastically three centuries could affect the spirit of Greek science.

Claudius Ptolemy, whose work was to be the final peak in the development of Greek astronomy, was entirely an 'astronomer' and not at all a 'physicist': he limited the scope of his work in the way in which the Babylonians had done before him. So far as mathematical astronomy went, Ptolemy summed up the work of his predecessors as Euclid had done earlier for geometry, adding a good deal of original material of his own. But, as we shall see, he was content to 'save appearances' and did not speculate about the underlying fabric of the heavens. He did not even care—as Eudoxos had cared—whether the mathematical constructions he used for astronomical purposes were consistent with one another; and the pattern he set was to be the shape of astronomy for centuries to come.

THE SCIENTIFIC BACKGROUND: AN ACQUISITION

While the direction of Greek history was permanently changed by the creation of Alexander's newly-conquered empire, Greek astronomy was similarly re-directed by the new knowledge which became available as a result of his success. With the capture of Babylon, Callisthenes was able to send home to Aristotle something of the greatest scientific interest: first-hand information about the accumulated records of the Babylonian astronomers. This was put to use straight away. Callippos, who was working with Aristotle, used some of the records in his own calculations of the length of the year and the month. (At first, only the astronomical records were sent back: the Babylonian methods of mathematical computation were transmitted more slowly.)

The Greeks were so impressed by the antiquity of this material that wild exaggerations began to circulate. Rumour reported that the Babylonian observations went back as much as 31,000 years! There was probably the same confusion here as obviously occurred in the Old Testament, where Mesopotamian kings such as Nebuchadnezzar were credited with fantastic longevity—months being read as years in the lists of their ages. Still, even 31,000 months would mean dating the earliest astronomical records at around 2800 B.C.; a date which is just possible since the chief cities of Babylonia were certainly flourishing by this time.

So, by 150 B.C., the tides of Babylonian and Greek astronomy had effectively flowed together. Partly owing to the volume and antiquity of these records, Greek astronomers now paid new attention to numerical precision. It became more and more difficult to fit the observed motions of the planets into Aristotle's tidy system of concentric spheres and the records revealed more and more aberrations in their paths. As time went on, the general framework of cosmological ideas became part of a religious picture; while astronomers concentrated independently on the mathematical techniques of computation.

Hipparchos, one of the greatest calculators, certainly cited the Babylonians' eclipse-records, and he probably also used their material when preparing his catalogue of stars. His own computational work was even more precise than theirs had been: for instance, he discovered the 'precession of the equinoxes', which they had overlooked. Having compiled a catalogue of more than 850 stars, with their celestial latitudes and longitudes, he compared their positions with those recorded by earlier astronomers and noticed that all the stars had apparently shifted together. The axis of rotation of the whole firmament of fixed stars seemed in this way to trace out a slow circle in the sky, completing one rotation every 26,000 years. (Explaining the observation in modern terms, we would infer that the Earth's own axis slowly changes its direction, rotating like the axis of a spinning top.)

To see just how exact Hipparchos was, consider his new estimates of the mean solar year and the mean lunar month. His value for the month—29 days, 12 hours, 44 minutes, 2½ seconds—differs from the figure accepted today by less than a second. For the year, he gave a value of comparable accuracy—365¼ days, less 5 minutes. This deficiency of a few minutes is compensated for in the modern calendar by skipping one leap year every 400 years: February 29th is due to be omitted from our calendar in the years A.D. 2000, 2400, and so on.

PTOLEMY'S MATHEMATICAL ASTRONOMY

Claudius Ptolemy lived from about A.D. 85 to 165. His astronomical masterpiece has traditionally been known as the *Almagest*. (This name itself is a significant historical relic, for it comes from a mediaeval Latin version of the title given to the book in eighth-century Baghdad: this in turn was an Arabic corruption of the Greek phrase for 'the greatest'—i.e. the largest of Ptolemy's treatises. The original Greek title might be translated as the *Mathematical Concordance of Astronomy*.) The greater part of the book expounds in detail geometrical devices which can be used to compute the ways in which the Sun, Moon, and planets move across the sky. The solar and lunar calculations of Hipparchos are taken over and expanded; and in addition Ptolemy gives a complete treatment of the motions of the five remaining planets. The task which the Babylonians had embarked on, using arithmetic, Ptolemy thus completed, using geometry.

The details of his mathematical methods are highly ingenious, but too complex for us to set out here. To understand the direction in which the astronomical argument later developed, we need bear in mind only his three principal geometrical devices: they are the *epicycle*, the *eccentric*, and the *equant*.

(i) The idea of the *epicycle* has already been mentioned in connection with Herakleides and Apollonios. The original purpose of the construction was to explain the 'retrograde motion' of the planets: this was done by supposing that the tracks of the planets consist, not of simple circles, but of a pair of circles added together—the centre of the smaller, quicker circular motion being carried around on the larger, slower circle at constant speed.

(ii) The idea of the *eccentric* had been used before Ptolemy's time by Hipparchos. Where, according to Apollonios, the planet moves on a quicker epicycle, we can instead make the *centre* of the deferent travel round the central Earth. The whole deferent circle then rotates off-centre like an eccentric gear-wheel.

(iii) The idea of the *equant* is one we have not previously met. It arose as follows. When Ptolemy tried to calculate the *speeds* at which the various heavenly bodies moved around their geometrical centres, he struck an unforeseen difficulty. Having constructed orbits for the Sun, Moon, and planets out of a combination of epicycles and eccentrics, he still found that the heavenly bodies were moving at an apparently non-uniform rate. He accounted for the most serious of these deviations by supposing that the planetary rotations were uniform, not as measured from the central Earth—nor even from the centres of their epicycles or eccentrics—but from a quite different point which became known as the 'equant': often, this was a point in space twice as far away from the Earth as the centre of the eccentric deferent.

His calculations gave results which in nearly every case squared closely with what one could observe using the simple astronomical instruments available at the time. Yet one thing about his methods strikes us nowadays as curious. He built up a new construction for the purposes of each *single* planetary calculation: he was content to show that the particular problem in question could be solved in this way. That done, he was prepared to put forward a new, and quite often an inconsistent, construction to deal with the next problem. For instance, to account for the *speed* of the Moon's motion he attributed to it a large epicycle, although this implied changes in the *apparent diameter* of the Moon far greater than we actually see. To deal with observed changes in the Moon's diameter he would use another construction. Nowhere did he provide a unique set of geometrical constructions capable of accounting for *all* the motions of a planet *at the same time*, let alone all the motions of all the planets, as Eudoxos had done. And this does not seem to have worried him. As an astronomer, he felt that his job was done once each separate 'appearance' had been 'saved'.

Whenever two constructions yielded equivalent mathematical results, there was, in Ptolemy's view, no astronomical difference between them. Given planetary observations displaying two independent kinds of irregularity, we are at liberty to account for the one anomaly in terms of 'eccentrics' and the other in terms of 'epicycles'—whichever way round we please. So far as Ptolemy is concerned, the question of the physical reality or unreality of these motions is beside the point.

Scientifically speaking, then, Ptolemy limited his aim. In general outline, he accepted the physics of Aristotle, but it played no part in his astronomy. The theory of the Heavens (he believed) was not a matter for physicists, but one for mathematicians.

In adopting this attitude, Ptolemy was simply turning his back on some crucial theoretical questions, which eventually had to be solved. From the physicist's point of view, the equant was a very unsatisfactory device. Mathematically it might do a job, but there was something intrinsically unreasonable about it. And in due course Copernicus made this feature of Ptolemy's astronomy the starting-point for his attack. As Copernicus argued: uniform circular motion could mean only one thing—namely, uniform circular motion. It could not mean motion circular around one centre, and uniform-in-speed around another. If one separated the centre of the orbit from the centre of uniform speed in this way, some physical explanation was surely called for—and Ptolemy gave none.

Yet, if all one wanted to do was to prepare tables of the positions of the heavenly bodies in the sky, there was not the same objection: all one needed was geometrical constructions which would *work*. Here Ptolemy was immensely successful: he was the first man ever to produce geometrical constructions which really fitted the astronomical records. Eudoxos had shown only in outline how one might eventually hope to account for the anomalies in the planetary motions: Ptolemy had actually shown in detail how one *could* do so. The geometrical methods he employed were in fact capable of infinite refinement, far beyond the point where he left them, and will give orbits of any shape: ellipses or even squares. (See the diagrams on pp. 141–2.)

Ptolemy, then, thought physics irrelevant to astronomy, and confined himself to mathematics. He justified this attitude in the introduction to his treatise. Here he explained the principles behind his method, and his point of view is strikingly different from that of the earlier Greek philosophers.

He begins by contrasting practical and theoretical knowledge:

> Sound philosophers have to my mind done well to distinguish the theoretical part of their subject from the practical part. For even where some prior theory does happen to have a practical application, great differences can nevertheless be found between the theory and the practice. In the first place, many people manage, at a practical level, to live perfectly good lives despite being uneducated; whereas one can get a grasp of general theory only through education. Furthermore, progress in theoretical and practical knowledge comes in different ways—in practical

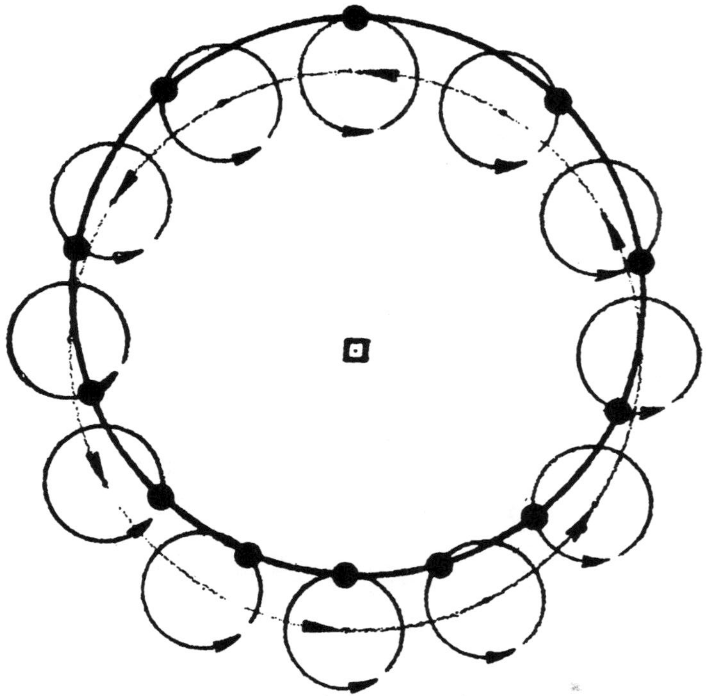

A

The versatility of epicycles

A, B and C show some of the various orbits that can be constructed by a combination of epicycles and deferents.

A shows how an epicycle can produce an effect equivalent to an eccentric. If the planet moves on the epicycle with such a speed that, half-way round the deferent, it has travelled once round the epicycle in the same direction, then the effect, as seen from the centre of the orbit, will be the same as if the planet had been moving along an eccentric circle

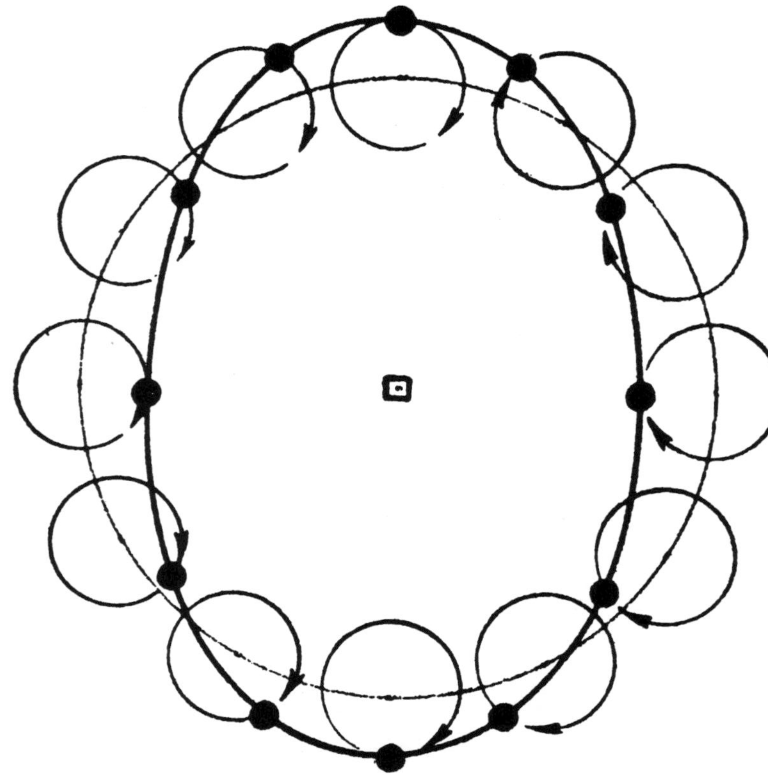

B

B shows how an elliptical orbit may be produced. In this case the path of the deferent is in an anti-clockwise direction, while the planet moves in a clockwise direction on the epicycle. This time, the planet travels half-way round the epicycle during the time it takes to go half-way round the deferent

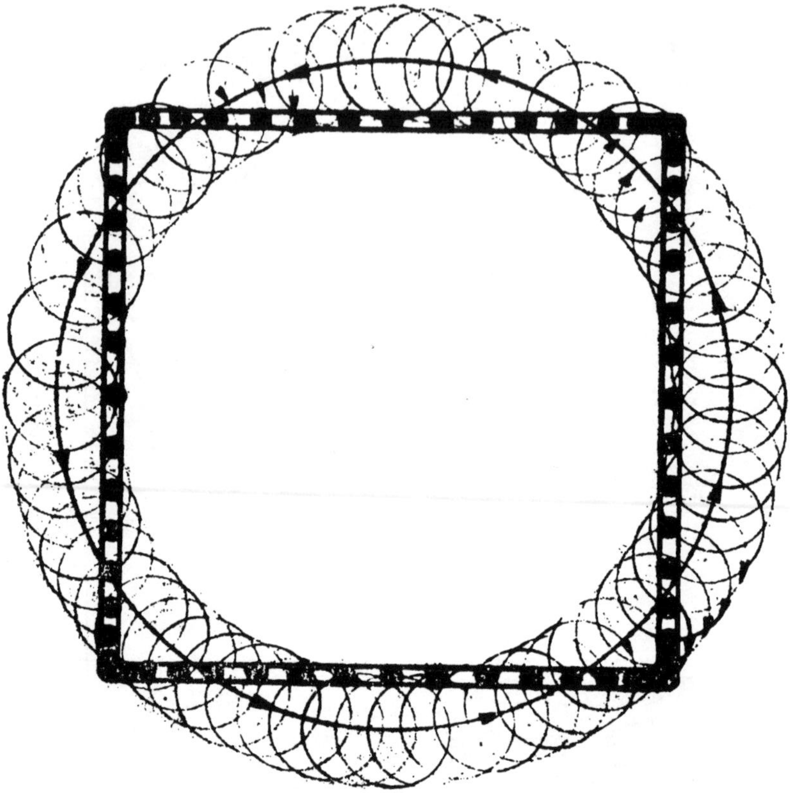

C

C demonstrates how, given a suitable choice of radii and speeds, one can get even an effectively square orbit. By appropriate variations in radii and speed, any kind of orbit likely to be met with in astronomy can be constructed

matters progress is made by continual repetition of acts of the same kind; but in matters of theory by the discovery of general rules or principles. Accordingly, in presenting our practical observations and operations, and even in reflecting on them, we thought it best to adopt an orderly exposition, so that our account should be elegant and clear, even when concerned with details; and in searching for the underlying principles of our subject—many and elegant as they are—to concentrate on those which are of the mathematical kind.

He then goes on to distinguish between the three principal branches of natural philosophy, and to indicate what, in his opinion, their interrelations are:

Aristotle very properly draws a further distinction, splitting up philosophical theory into three immediate subdivisions: physical, mathematical and theological. Things of all sorts can be analysed into matter, form and change—not that any one of these things can really be regarded as having any existence independently of the others, except in thought. (i) Thus, anyone whose primary concern is with the ultimate source of all *change* in the universe will find himself in the end led back to the unobservable and unchanging God; and, the nature of God being a theological matter, we can conceive of his activity only as taking place far above us, among the highest things in the universe, far away from the objects we directly observe with our senses. (ii) On the other hand, the nature of ordinary *material* things, with their continually changing properties—their colour, heat, taste, texture and the rest—is a matter for physics, things of this sort belonging for the most part to the sublunary impermanent realm.

There remain (iii) changes in *form*: i.e. in trajectory, shape, quantity, size, position, time and the rest. These things, which are our present concern, are the subject-matter of mathematics and occupy an intermediate position between the other sorts of things. . . .

Evidently, two of these three different sorts of theory would have to be expounded in a speculative rather than a scientific manner: (i) theology, since its objects are both unobservable and hard to grasp, and (ii) physics, because the material things which are its concern are so unstable and difficult to fathom that one can never hope to get philosophers to agree about them. The systematic study of *mathematical* theory alone can yield its practitioners solid conclusions, free of doubt, since its demonstrations—whether arithmetical or geometrical—are carried out in a manner that admits of no dispute.

This, then, was the sort of theory we decided we should study for preference and to the best of our ability: concentrating on its application to things of a divine and a celestial kind. The science of celestial things alone being concerned with eternal and unchanging objects, we thought it possible in this way to arrive at conclusions which were both clear and orderly, and would also have an eternal validity, as the conclusions of a true science should have.

Compare Ptolemy's position, as stated here, with the attitudes of the earlier Greek natural philosophers. We are half-way back to the Babylonians. Theology is concerned with 'the highest things in the universe', i.e. with the things in the uppermost divine Heavens: physics, on the other hand, is concerned primarily with the changeable things of the Earth. In either case, we can employ mathematical methods to work out—by extrapolation from past experience—the way in which things can be expected to change. But when it comes to arguments about the actual nature of things in the Heavens and on the Earth, this is something for separate disciplines. The original Greek ambition to explain heavenly happenings in terms of causes familiar to us on the Earth has been abandoned.

Again: compared with mathematics, physics will inevitably be a speculative science, about which we cannot hope to get general agreement. The different schools of physicists will go on arguing for ever without convincing one another, any more than opposing theologians. The full consequences of this change of attitude will become clear shortly. In effect, it put an end to the old faith that human reason could discover the natural principles governing the behaviour of things.

At first sight, Ptolemy seems to be following Plato, for whom true knowledge in science could come only as a result of intellectual insight into the mathematical order of Nature. The Heavens were neither entirely perfect nor unchanging, but they were as near changeless as material things could be; so that, having reflected on the mathematical orderliness of the planetary system, a man might desire to establish a similar harmony in his own soul. Indeed, Ptolemy sees in the study of astronomy some of the same morals as Plato:

> It will help, more than anything else, to make us better people, by concentrating our attentions on the beauty and value of moral conduct. For the correspondence we find between the order of divine celestial things and the order of our mathematical propositions encourages students of mathematical astronomy in their love of that divine beauty, and in this way accustoms them to using it as a model for their own behaviour, assimilating—as it were—the powers of the soul to it.

But there is a crucial change of tone. Ptolemy is nearer in his attitude to some of the neo-Platonists he would have known in Alexandria. In their view, complete 'insight' into eternal things should not only give one an intellectual grasp of Nature, but also place one in mystical union with the Divine. Rational science *alone* was no longer relied on to give one true knowledge: a full understanding of the Heavens, it was thought, came only through personal acquaintance with the Divinity. Aristotle's *physical* distinction between the changeable Earth and the changeless Heavens was now taken with full *theological* earnestness; things in the Heavens were once again made objects of worship, as they had been in Babylonian times; and astronomy was once more valued for its connection with astrology, not with physics.

THE WIDER REVOLT AGAINST PHILOSOPHY

By the time we reach Ptolemy, then, the original rational impulse behind Greek inquiries into Nature has evaporated. The roots that science had struck in classical Greece were never deep, and most people remained unaffected by the work of the natural philosophers. Astronomy was the one science which might have made a considerable impact on contemporary common sense; but now it seemed to have failed. Had there been a really satisfactory theory of the workings of the planetary system, the vanity of astrology might have been universally recognized. Instead, the 'astral religion' which philosophers had originally set out to displace kept its hold. Further, when the focus of scientific activity moved to Alexandria, it passed to a region where astrology had a long and proud tradition. Where Athenians at large could be indifferent to the rival claims of astronomy and astrology, most Egyptians naturally sided with the astrologers.

So, from about 100 B.C., philosophy was on the defensive: one finds a growing preoccupation with questions of a technological kind, on the one hand, and with questions of a religious kind on the other. Between the two, rational scientific speculation gradually got squeezed out. Greece and Egypt were both absorbed into the Roman Empire, and the Romans were primarily interested in practical affairs, rather than in questions of theory. Later Greek philosophers moved away from rational speculation about Nature, and became preoccupied rather with questions of morals and theology.

The Stoic School, founded by Zeno of Citium—no relation of the paradoxical Zeno—had a great attraction for the Romans, and won the respectful interest of the author, Cicero. The Stoics had some valuable scientific ideas, particularly in connection with matter-theory, but for many of them the Divinity of the Heavens—which for Aristotle was a theoretical insight—was important rather as a profound religious truth. On this basis, some of them even built up a sophisticated kind of star-worship, teaching that a man's soul escaped at death from his body, to be reunited with his own personal star. (Plate 5.) They believed that all natural events were causally determined, but this belief encouraged not scientific enquiry so much as a faith in divination.

Among the Romans, the serious alternative to Stoicism was the philosophy of Epicurus. This doctrine did no more than Stoicism to encourage scientific work: if anything, the Epicureans were even less interested in questions of astronomy. They turned men's attention right away from the Heavens, arguing that what went on in the sky was of no concern to men, whose proper business was with the problems of life on this Earth. The Roman poet Lucretius, who popularized Epicurus' views in the first century B.C., even dismissed the idea of the Antipodes and treated the sphericity of the Earth—which had been a commonplace in Athens for several centuries—as an entirely unproved speculation.

So the failure to resolve the fundamental problems of scientific astronomy had very wide repercussions. Even Ptolemy followed the Stoic doctrine in its main points. In fact, he wrote a companion volume to the *Almagest* in four parts, known as the *Tetrabiblos*, which did for astrology what he had already done for mathematical astronomy.

Ptolemy points out in the preface the obvious influence of the heavenly bodies on terrestrial affairs, and justifies astrology by comparing it with weather-prediction. Any fool (he implies) can see that the Sun and Moon affect the growth of crops, the rise and fall of the seas, the changes in the weather, the time of germination of seeds, and so on:

> Suppose, then, a man knows accurately the movements of all the stars, the Sun and the Moon, and overlooks neither the place nor the time of any of their conjunctions . . .; and from these data is able to work out, both by calculation and by successful conjecture, the distinctive effects which will

result from the combined operation of all these factors: what is to prevent him from telling how the atmosphere will be affected by the interaction of these phenomena on any particular occasion—e.g. that it will be warmer, or wetter? Why should he not, in the same way, by considering the nature of the astronomical environment at the time of birth, work out for any individual man the general character which his temperament will have, e.g. that he will have such-and-such bodily or mental characteristics?

Astrology, he argues, is not only possible but useful. We all use our knowledge of the Heavens and the seasons to determine the best times for sowing, the favourable occasions for setting sail, the times to lay in firewood, and so on: 'No one ever condemns such practices as these either as being impossible or as being useless.' Why then should we regard astrology as useless? He accordingly expounds the political significance of eclipses and the astrological powers of the different planets: which of them bring warmth or moisture, which of them are beneficent or maleficent. In every way, Ptolemy regarded astrology in the same light as tide-prediction or meteorology; and one finds Kepler still tolerating it as late as A.D. 1600. Until Newton established a theory of the *forces* by which heavenly bodies can act on earthly things, it was difficult to demonstrate the scientific worthlessness of astrology.

The spirit of rational inquiry was not entirely dead, but the prospect of a short-cut to Truth naturally discouraged rational speculation about Nature. Even philosophy had now to justify itself as the servant of religion, and was studied, not as an end in itself, but rather to promote religious piety:

> A sincere love of philosophy consists uniquely in the desire to know the Divinity better, by habits of contemplation and holy piety. Many people have already adulterated it with all kinds of false wisdom . . . mixing it up with unintelligible sciences: arithmetic, fractions, geometry and the rest. Pure philosophy, which depends solely on piety towards God, should interest itself in other sciences only in so far as these . . . encourage one to admire, adore and bless the craftsmanship and intelligence of God. . . . To love God with a simple heart and soul, to revere the works of God, and finally to show in one's life thankfulness for the Divine will which alone is the fullness of the Good: that is the philosophy which is unspotted by any harmful curiosity of mind.

The idea that the rational curiosity was harmful became the slogan for a positive counter-attack against Greek science and philosophy. In Ptolemy's own century, the theologian Tertullian put the idea forcibly:

> What has Jerusalem to do with Athens, the Church with the Academy, the Christian with the heretic? Our own doctrine comes from the House of Solomon, and Solomon himself has taught us: one must seek for the Lord in the simplicity of one's heart. Let us have nothing to do with a Stoic Christianity, or a Platonist or dialectical Christianity. All curiosity is at an end after Jesus, all research after the Gospel. Let us have Faith, and wish for nothing more.

By the fourth century A.D., conditions for scientific work in Alexandria were deteriorating rapidly: work of great originality was being done in mathematics alone. The critical spirit of science was swept aside, and the city was the scene of religious riots. According to one tradition a mob of Christian fanatics set fire to the Library: according to another, this was the work of Arab conquerors three centuries later. Whoever did it destroyed a greater part of our intellectual heritage than anyone before or since. By the sixth century A.D. one had either to be converted to Christianity or leave.

There was a brief revival around A.D. 500, in which the best-known figures are two bitter intellectual enemies, Philoponos and Simplicius. After his years of scientific work, Philoponos (whose arguments about dynamics we looked at in the last chapter) became a Christian, and engaged in theological debates. Ironically enough, the only lasting outcome of this theological enthusiasm was his posthumous condemnation as a heretic. Simplicius left Alexandria for Athens: but the Emperor Justinian suppressed the Academy in the year A.D. 529 after nine centuries of work. With six other leading philosophers, Simplicius moved on to Persia, hoping to find a more enlightened atmosphere. They eventually returned to the Greek world, but the pattern of the future was set. From now on the general direction of intellectual movement was eastward, and Christendom generally settled down into the Dark Ages.

In retrospect, one can see that the decline of Greek science did not come about through any loss of individual competence: the few men who carried on the scientific tradition of the Greeks were as able as ever. Philoponos and Simplicius both of them displayed a complete grasp of the issues with which Aristotle and Plato had been concerned. Quite apart from his anticipation of the idea of momentum, Philoponos made several other original suggestions: he even rejected Aristotle's fundamental doctrine that the Heavens are immutable and are composed of a material quite distinct from that of earthly things. (This prompted an acid enquiry from his unconverted opponent, Simplicius, as to how he reconciled *that* view with his new-found religion.) What brought science in the Greek world to a halt was, rather, a general deterioration in the conditions of scientific work, and a widespread scepticism about the value of a rational natural philosophy. So, however talented they might be, it became impossible for men like Philoponos and Simplicius to pass the scientific tradition on to later generations. In Christendom, at any rate, physics had been quietly anaesthetized.

NOTE: ARCHIMEDES AND THE CIRCLE

The Greeks asked: 'What straight-sided figure has the same area as a given circle?' This was part of the general problem of 'squaring the circle'. Archimedes answered this question by an argument involving 'limits', and so produced a formula for the area of a circle equivalent to our own πr^2. His proof can be presented as follows:

Any right-angled triangle can be regarded as one half of a rectangle whose sides, a and b, form the shorter sides of the triangle and whose diagonal, c, is the hypotenuse. The area of the triangle will accordingly be $\frac{1}{2}ab$.

We can now build up a series of figures out of right-angled triangles alone, which approach in area to a given circle as closely as we please. We can start by drawing a square with corners on the circle: then double the number of sides, getting an octagon; do the same again to get a 16-gon; and so on. Each of these figures will be smaller in area than the circle, but we can always reduce the deficiency by increasing the number of sides further. To 'exhaust' the

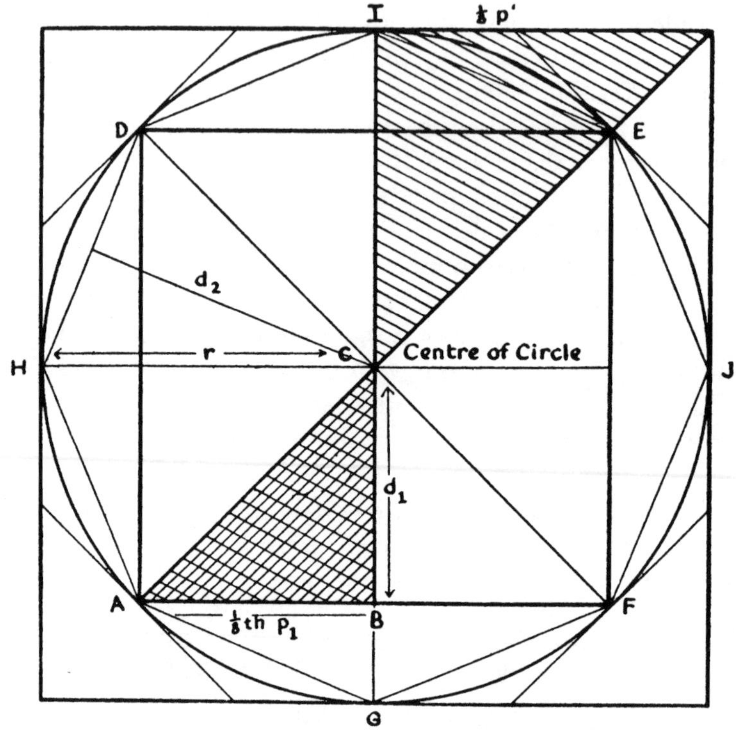

Archimedes' method for finding the area of a circle

Triangle ACB will have area $\frac{1}{2}d_1 \times \frac{1}{8}p_1$
Total area of Square ADEF will be $\frac{1}{2}\, d_1\, p_1$
The Octagon HDIEJFGA will have an area $\frac{1}{2}d_2\, p_2$, and so on
d is always the length of the perpendicular from the centre of a figure to any side of the polygon and p is the perimeter

area of the circle completely, we would have to increase the number of sides indefinitely: in this way, the circle can be thought of as the 'mathematical limit' which we approach as the sides of the inscribed polygon increase in number.

Now the square is composed of eight equal right-angled triangles (of which one sample is hatched in the figure), each of them having a radius of the circle as its hypotenuse. If the perpendicular from the centre of the circle to one side of the square has length d_1

and the perimeter of the square (i.e. the total length of its sides) is p_1, each triangle will accordingly have an area $\frac{1}{2}d_1 \times \frac{1}{8}p_1$; and the total area of the square will be $\frac{1}{2}d_1p_1$. Similarly, the octagon will have area $\frac{1}{2}d_2p_2$, the 16-gon area $\frac{1}{2}d_3p_3$, and so on; where in each case d is the length of a perpendicular from the centre to any side of the polygon and p is its perimeter.

As the number of sides increases, d gets closer and closer in length to the radius of the circle, while p approaches continuously closer to the circumference. The areas of the polygons thus approach, without ever quite reaching, the value $\frac{1}{2}$ radius $\times$ circumference.

Now build up a second series of figures *outside* the circle. A square of perimeter p' will have area $\frac{1}{2}rp'$; an octagon will have area $\frac{1}{2}rp''$, a 16-gon area $\frac{1}{2}rp'''$, and so on. The areas of these figures will all be larger than the area of the circle, but we can once more cut down the excess as much as we please by increasing the number of sides sufficiently. As we gradually 'shrink' the circumscribed polygon on to the circle in this way, the area will again approach closer and closer to the limiting value of $\frac{1}{2}$ radius $\times$ circumference.

All the inscribed polygons are smaller than the given circle, and all the circumscribed ones are larger; but they converge to the same limit. This limit, Archimedes argued, must represent the area of the circle. Since the number π is defined as the ratio of the circumference of a circle to its diameter, we can substitute the value $2\pi r$ for the circumference in Archimedes' expression for the area, '$\frac{1}{2}$ radius $\times$ circumference', to get the more familiar formula

$$\text{Area of a circle} = \pi r^2$$

This argument contains the germ of the idea of 'infinitesimals'. Later on, the method of splitting up figures or magnitudes into smaller and smaller parts, and obtaining a required formula as a 'limit', was carried over from geometry and statics (where Archimedes introduced it) into dynamics. This led, at the end of the seventeenth century, to the introduction of the calculus of 'fluxions' or 'infinitesimals' by Newton and Leibniz, and so to dynamics as we know it.

The later period in Greek science has been neglected until recent years, both by scholars and by popular writers. There is a brief account of Alexandrian science in

M. Clagett: *Greek Science in Antiquity*

The Ptolemaic system of astronomy is discussed at some length in Dreyer's book, and also more recently in an appendix to the second edition of

O. Neugebauer: *The Exact Sciences in Antiquity*

For Greek natural philosophy in the centuries after Aristotle, consult

S. Sambursky: *The Physics of the Stoics*
G. S. Kirk and J. Raven: *Stoics and Epicureans*
A. J. Festugière: *Epicurus and his Gods*

Heath's book on *Aristarchus* discusses the development of Greek astronomy in the period before Ptolemy, but no acceptable edition or translation of Ptolemy's *Almagest* exists in English at present (1960). S. Sambursky has in preparation a long-needed collection of scientific texts dating from the period under discussion in this chapter.

For late Alexandrian theories, and especially the influence of the neo-Platonists, it is still necessary to consult the scholarly treatises: e.g.

A. J. Festugière: *La Révélation d'Hermès Trismegiste*

For the versatility of Ptolemy's epicyclic constructions, see the paper by N. R. Hanson in *Isis* (Seattle) for 1960. The use of epicycles to explain retrograde motion is demonstrated in the Greek section of the film *Earth and Sky*.

Part II: The New Perspective and its Consequences

6

The Interregnum

AFTER Ptolemy's *Almagest* no important new ideas found a place in astronomy for nearly 1400 years. In the sixteenth century A.D., astronomers were still concerned with the very same problems that Ptolemy had tackled, and were only just beginning to see a way beyond. To many people, this period of stagnation appears mysteriously, even inexcusably long, and they look for someone to blame. Surely (they feel) the interruption would have ended sooner if men had not been prevented—by the Church perhaps—from recognizing truths which were there to see.

To think in such terms is to miss two crucial points. The change-over to our modern view of the universe was not a matter of acknowledging obvious facts. It called for a series of daring intellectual steps, which could be taken only as new evidence gradually accumulated and in the face of weighty difficulties and objections. This inevitably took time. In any case, the process could not begin at once. The young growth of Greek science, having been cut down in Athens and Alexandria, had to be transplanted and grown again in fresh soil. Many years passed before it could root itself in a new environment firmly enough, and for long enough, to set fresh fruit.

THE ROUNDABOUT JOURNEY

In late antiquity the intellectuals of Alexandria became interested in theology and the occult sciences rather than in philosophy and the natural sciences. In Athens the Academy was closed. In Constantinople the authorities of the mutilated Eastern Roman Empire were preoccupied with questions of politics and theology. Something was known, of course, about the ancient tradition of natural philosophy, but there was no serious opportunity to carry it further. The books and manuscripts remained in the libraries, gathering dust.

In the eyes of the Byzantine authorities, their state was the political embodiment of Christianity, much as, to the Russian authorities today, the U.S.S.R. is the political embodiment of Marxism. This identification of the Christian Church with the Byzantine State tended to turn all intellectual debates into ideological ones: opinions were judged, not only on their merits, but also by their supposed political tendencies. Unorthodox intellectual views ran the risk of condemnation, not only as heretical but also as treasonable. The resulting intellectual atmosphere hardly encouraged the imaginative speculation needed if science was to make further progress. So natural philosophy was put in the deep-freeze.

Greek science was in any case associated in men's minds with paganism. After A.D. 500., few of the Church fathers even bothered to condemn it: only Lactantius, in the early fourth century, made any serious attempt to challenge the Greek arguments, dismissing the idea that the Earth was a sphere and ridiculing the Antipodes as entirely impossible. In a popular sixth-century geography by Cosmas, the Greek astronomers were also mocked:

> . . . with supercilious airs, as if they far surpassed in wisdom the rest of mankind, they attribute to the heavens a spherical figure and a circular motion, and by geometrical methods and calculations applied to the heavenly bodies, as well as by the abuse of words and by worldly craft, endeavour to grasp the position and figure of the world by means of the solar and lunar eclipses, leading others into error while they are in error themselves in maintaining that such phenomena could not present themselves if the figure were other than spherical.

Around A.D. 1100, the historian Anna Comnena could speak with respect of Ptolemy, but the Empress Irene in the thirteenth century laughed at the scholar Acropolita for saying that eclipses were caused by the Moon coming between the Earth and the Sun. For most of the theologians, a study of nature was interesting only as a source of allegories, which were used to illustrate the virtues and beauties of the spiritual life. In this way, science simply got bypassed.

As for the Western Roman Empire, from the time of the Germanic and Frankish invasions right up to A.D. 1000, its contribution to the scientific tradition was entirely negligible. Economically, these invasions reduced it to a subsistence level: the currency was debased and the opportunities for intellectual life were very limited. There, if anywhere, the Dark Ages were really dark.

As things turned out, therefore, the emaciated relics of Greek science travelled to mediaeval Europe chiefly by a roundabout route: they migrated several times from one intellectual centre to another in the Middle East, and finally re-entered Europe through the Arabic kingdoms of Spain and Sicily. So the scientific works of men like Aristotle and Ptolemy reached the budding universities of France, Italy, and England at third- or fourth-hand. In some cases they had been translated as many as three times. Digested and loaded with comments, these texts arrived in Europe to be forcibly joined together—like the incomplete fragments of four or five separate jig-saw puzzles—in the picture known to the Middle Ages as the World-System of Antiquity. (Plate 9.)

Only the chief milestones in this roundabout journey need be mentioned at this stage. It was no accident that Simplicius and his companions turned first to Persia in hope of a more sympathetic reception. The monasteries in the Eastern Byzantine provinces of Armenia and Syria were already centres of scholarship, where Greek philosophy and science were being translated into Syriac—a Semitic language, related to Hebrew and Arabic—and their works spread across the frontier into Persia. The doctors and teachers at the great hospital set up by the Persian king in Jundishapur, near the border between modern Iran and Iraq, were largely Syriac-speaking Jews and Christians, and in due course the place became a considerable centre of learning.

Early in the seventh century the followers of Mahomet, coming up from Arabia, overran the whole Middle East: in a few decades they had occupied the whole of Persia, Egypt, the North African coast, and Southern Spain. But the conditions for intellectual life remained good, owing to the religious tolerance of early Islam. From A.D. 750 onwards the great patrons of science were the Caliphs of Baghdad, the splendid new city a few miles upstream from the ruins of Babylon. It is said that their interest in astronomy was first aroused by a visitor from India. Knowledge of the science had reached India as a result of Alexander's conquests, and the existing Hindu doctrines about astronomy represented something between Hipparchos and Ptolemy. The Caliph Al-Mansur ordered these doctrines to be translated into Arabic and adapted for local use. Later, a translation of Ptolemy's *Almagest* was made, and a new observatory was built at Baghdad in A.D. 829, at which the most noted of the Arabic astronomers worked. At the new Arabic University in Baghdad, the Bait Al-Hikma, the works of Aristotle, Archimedes, and other scientists were also translated into Arabic, some from Syriac, others directly from the Greek.

So, for two centuries at least, Baghdad was the great centre of mathematical and astronomical study. Religious tolerance and official patronage kept alive in the Islamic world an active interest in Nature. Baghdad succeeded Athens, Alexandria, and Jundishapur as the centre of attraction for scholars: many of the leading Arabic scientists were 'Arabic' only in language, coming from distant parts of the Empire. From Cordova in Spain to Samarkand in Central Asia, the Islamic world was a cultural unity.

The golden age of science in Baghdad lasted little beyond A.D. 1000. In medicine and chemistry the Arabs made substantial contributions, but in astronomy and dynamics they had time only to digest and adapt the existing traditions. They developed and greatly improved the astronomical instruments left to them by the Greeks: notably, the astrolabe, an ingenious kind of sextant-cum-computer which was widely used for desert navigation, as well as for astronomy. (Throughout the centuries, caravans have moved chiefly by night, to avoid the searing heat of the midday sun—so the stars were a natural guide.) They made fresh estimates of the size of the Earth and the relative distances of the planets, and investigated minor irregularities in the planetary movements, but made no fundamental changes in Ptolemy's methods. Certainly they never questioned the central features of his picture: at most, they tried to bring it more closely into line with the ideas of Aristotle. The problem was to reconstruct Aristotle's system of solid, transparent spheres so as to make mechanical sense of Ptolemy's geometrical constructions, as

Aristotle himself had done for the geometry of Eudoxos. The resulting solid, mechanical contrivance, a wonderfully complicated hotch-potch of eccentric transparent spheres separated by epicyclic ball-bearings, was later to become an essential part of the cosmology of Mediaeval Europe. The 'intelligences' which some of the Arabs supposed necessary to keep the planetary spheres in motion then became identified in the popular mind with 'angelic powers'.

After about A.D. 1000, the centre of intellectual activity in Islam shifted to the West, to Spain, and scholars in the neighbouring parts of Europe had their appetites whetted for the lost treasures of ancient science. Wars in the twelfth and thirteenth centuries were not the all-out things they have since become: though, in Spain, Christian Castile and Islamic Andalusia were in a continual state of political and religious strife, there was a steady flow of scholars and translators across the shifting frontier. Around 1150, the scholar Gerard of Cremona seems even to have set up a kind of translating-factory at Toledo, turning out Latin versions of Arabic texts on a great range of subjects. A similar traffic in ideas was established in Italy, linking Islamic Sicily with the new hospital at Salerno near Naples.

From the twelfth century on, the political power of Islam began to decline. In its greatest days, there had been no hostility between philosophers and theologians. But now, under the growing political threat, the pattern we saw earlier in Alexandria began to reappear: rational enquiries in science and philosophy were condemned for the first time as corrupting the truths of Islam. The scientific tradition in the Arab world began to dwindle, and if their knowledge had not been taken up into Europe when it was, the mainstream of astronomy and physics might have moved even further East, into Asia.

The first great master in the Latin revival was the scholar Gerbert, who later became Pope Sylvester II. From A.D. 972 he was an active student and teacher of logic, mathematics, and astronomy at Rheims. His own knowledge was limited to the scanty material already available in Latin, but his pupils began to look further afield: shortly after A.D. 1000 the first serious trickle of learned material began to cross the frontier from Moslem Spain. Fulbert, Bishop of Chartres from 1006 to 1028, founded the first of the great cathedral schools, and from this time on the essential chain from master to pupil was re-established in Europe, continuing unbroken to the present day.

Fulbert himself was quick to take advantage of the new intellectual contacts with the Arabs. The astrolabe reached Europe from the Islamic world in his time, and rough astronomical measurements began to be made once again. Hitherto, learning had been valued in the Latin world for the improvement it could give to man's soul: Fulbert encouraged learning of all sorts—for its own sake. The Crusades led incidentally to a new interest in Greek, and the Venetian sack of Constantinople (the only tangible outcome of the Fourth Crusade) created a trans-Mediterranean trade in old manuscripts comparable to the trans-Atlantic rare book trade of today. By the fifteenth century, old Greek manuscripts had acquired a snob value among the rich and educated, to which Robert Browning alludes in the poem *The Bishop Orders his Tomb*. The bishop fears that his sons will economize on his tomb and tries to bribe them with promises—to be fulfilled from beyond the grave—of the most desired and desirable possessions:

> 'Tis jasper ye stand pledged to, lest I grieve
> My bath must needs be left behind, alas!
> One block, pure green as a pistachio-nut,
> There's plenty jasper somewhere in the world—
> And have I not St Praxed's ear to pray
> Horses for ye, and brown Greek manuscripts,
> And mistresses with great smooth marbly limbs?

THE MEDIAEVAL REVIVAL

It is hard to realize just how ignorant the learned men of Western Europe were in the year A.D. 1000, and how much leeway had to be made up before they could compete with the Arabs, let alone improve on Aristotle. At the time of the First Crusade, Europe was a backwater. Constantinople was ten times the size of any city in Western Europe, and the authorities there understandably treated the self-invited Crusaders as an uncivilized rabble. Cordova in Arab Spain equally out-shone the European capitals. It had a library of half-a-million volumes at a time when the Royal Library at Paris contained perhaps two thousand.

Cut off from all but fragments of the ancient tradition, European scholars could not even understand the elementary texts which they still possessed. One of Fulbert's most famous pupils, Reginbald of Cologne, found a remark in Boethius which neither he nor Ralph of Liège could make out. It referred to the elementary theorem in Euclid, that 'the interior angles of a triangle are equal to two right angles'. Neither of these men, who were among the leading scholars of Western Europe, had heard of the theorem before or even knew what the phrase 'interior angles of a triangle' meant. Their guesses were wide of the mark: Reginbald, for instance, thought it meant 'the right angles on either side of a perpendicular dropped from the apex of a triangle to the base'. In rebuilding the scientific tradition after A.D. 1000, therefore, men in Western Europe had in effect to start from scratch.

The situation being what it was, some of the manifest distortions in their resulting picture of ancient science are understandable. Few mediaeval scholars could read Greek, and the journey through Syriac and Arabic—like the party game of Gossip—introduced into the texts not only some additional truths, but also a fair number of corruptions and misinterpretations. The further task of translating the ancient texts into Latin had to be carried out by men unfamiliar with the technical terms and ideas involved, and this naturally added to the confusion. When first-hand texts began to arrive direct from Constantinople, the task became somewhat easier; but, to begin with, there was no way of distinguishing the mistakes of (say) Aristotle himself from the blunders of translators and interpreters. Since in many cases Aristotle's personal knowledge and insight had been so deep and exact, we need not wonder that European scholars soon came to feel their intellectual inferiority acutely; it was only reasonable to hesitate before concluding that Aristotle had in fact been wrong.

Still further confusion resulted from the tendency of mediaeval scholars to telescope the astronomical doctrines of Eudoxos, Aristotle, and Ptolemy, and to squeeze out all disagreements and inconsistencies. Looking back after a whole millennium they could not reconstruct the historical development of Greek thought in perspective. (Despite all our intellectual advantages, we ourselves are often guilty of the same error, in looking for a single consistent 'mediaeval world-picture', embracing everything from the year 1050 up to the time of Galileo.)

The reconstructed picture differed from the original Greek theories, not only in perspective, but also in emphasis. Lacking a direct tradition of teaching, the mediaeval scholars had to place their own interpretations on the different parts of their picture. For instance, we find Aristotle's ideas of the 'unmoved mover' and the outermost Heaven given a religious slant strikingly unlike the one which Aristotle himself had intended. The new emphasis on the relations between natural philosophy and theology altered the whole perspective of the cosmological picture.

A crucial part in the mediaeval revival of science was, inevitably, played by the Church. Some scholars have trumpeted against scholasticism and denounced the influence of the Church as 'reactionary'—reading back into the high Middle Ages the ideological panic shown by the Inquisition in the years around 1600, of which the trial of Galileo was one symptom. Actually, the Middle Ages were for the most part a period of growing intellectual confidence, and the Church was happy to provide the institutions needed for the rehabilitation of philosophy. In the teaching orders especially, men were given an opportunity to build up a body of learning and pass it on to generations of pupils in unbroken succession, in a way that had not been possible for hundreds of years. During the Middle Ages, scholars could once again devote themselves to the mastery of technicalities and specialize in subjects having no immediate practical importance. Without this re-creation of a tradition of disinterested theorizing, science could never have been reborn in Europe as it was. (Plates 7 and 8.)

There were, of course, set-backs. The *Physics* and *Metaphysics* of Aristotle came to Europe through Averroes, whose commentary emphasized the difficulty of reconciling Aristotle's science with orthodox theology, whether Muslim or Christian. As a result, there were a few decades, from 1210 on, when these works were barred from the curriculum of the schools. But Thomas Aquinas (1225–

1274) made it his task to show that nothing essential in Aristotle's system was incompatible with orthodox Christian theology; and—though it was not easy—integrated the main features of the Aristotelian synthesis into the orthodox conception of the universe. From this time on, the great system of theory which Aristotle had left became the foundation of natural science in the West.

By 1600, the commitment of the Church to Aristotle had become an embarrassment. It was difficult to carry through the reconstruction that physics by then required, without coming into conflict with the theological authorities. The Inquisition lapsed into practices of intellectual repression that the Church would not have contemplated a century earlier. (Copernicus' book was, in fact, put on the Index for the first time sixty years after it appeared.) But this was 500 years after the first serious revival of scholarship. One will not get a just view of the contribution which the Church, as an institution, made to the development of scientific ideas during the Middle Ages, by looking only at its period of anxiety: say, from 1550 on. One must look also at the years of confidence—from 1000 to 1500—when its leading scholars and doctors were mastering the science of the ancients, and making their own first cautious moves forward. Against this background alone can the intellectual tasks facing Copernicus and his successors be properly appreciated and understood.

THE BACKGROUND TO COPERNICUS

Sixteenth-century astronomers inherited from the Middle Ages two distinct traditions: on the one hand, a set of computative techniques, which had remained effectively unchanged from Ptolemy's time; on the other, a general cosmological picture. This last was an amalgam of Aristotle's physics and Ptolemy's epicycles, embedded in a setting constructed by neo-Platonist mystics and Christian theologians.

Toledo was the centre from which Gerard of Cremona circulated his translations of the classics of Arabic science, and it was here also that in 1252 a new set of planetary tables was completed, using Ptolemy's methods. The tables were computed by a team of

Jewish and Christian astronomers, with the active interest—and possibly also the collaboration—of King Alfonso X of Castille. (Alfonso was evidently appalled by the complexity of the task: if God had consulted him when planning the Creation, he said, he could have told Him a thing or two!) These *Alfonsine Tables* circulated throughout Europe for 300 years, and during that time were the best available. They were put into print in 1483, and were superseded only in 1551, when Erasmus Reinhold prepared his *Prutenic Tables* for the Duke of Prussia, on the basis of Copernicus' new techniques. Even so, the later tables were not really more reliable than the earlier ones: they were just more up-to-date.

The cosmological picture played no part in professional treatises on astronomy. One finds it, rather, described in popular handbooks of cosmology: for instance, in John of Holywood's *De Sphaera*, which, from the middle of the thirteenth century, remained for three centuries a standard elementary textbook. The same picture reappears in the poetry of the time, and even in the most sophisticated treatises of theology, such as Aquinas' *Summa Theologiae*. There, the unchangeable region outside the sphere of fixed stars was identified with the Heaven of the Christian religion. The nine orders of angels were allotted different functions, and were associated with different parts of the cosmos: three grades operated in the outermost empyrean, three on the Earth itself, and three in the intervening region. The humblest, most imperfect creation was to be found on the Earth; as one travelled outwards through the heavenly spheres, the successive heavens were progressively more perfect; while beyond the outermost sphere there lay the habitation of God and all the elect.

In Dante's great three-part poem, *The Divine Comedy*, this symbolism is carried even further. Hell is located in the centre of the Earth, so that man's life is—symbolically—lived out midway between the perfect, superlunary world, and the utterly corrupt infernal regions. Further parallels (to be examined in a later book) brought the central ideas of zoology and matter-theory into line with this celestial hierarchy. All of natural science seemed, accordingly, to confirm the central insight of religion: that the universe was (so to speak) a golden apple with a rotten core.

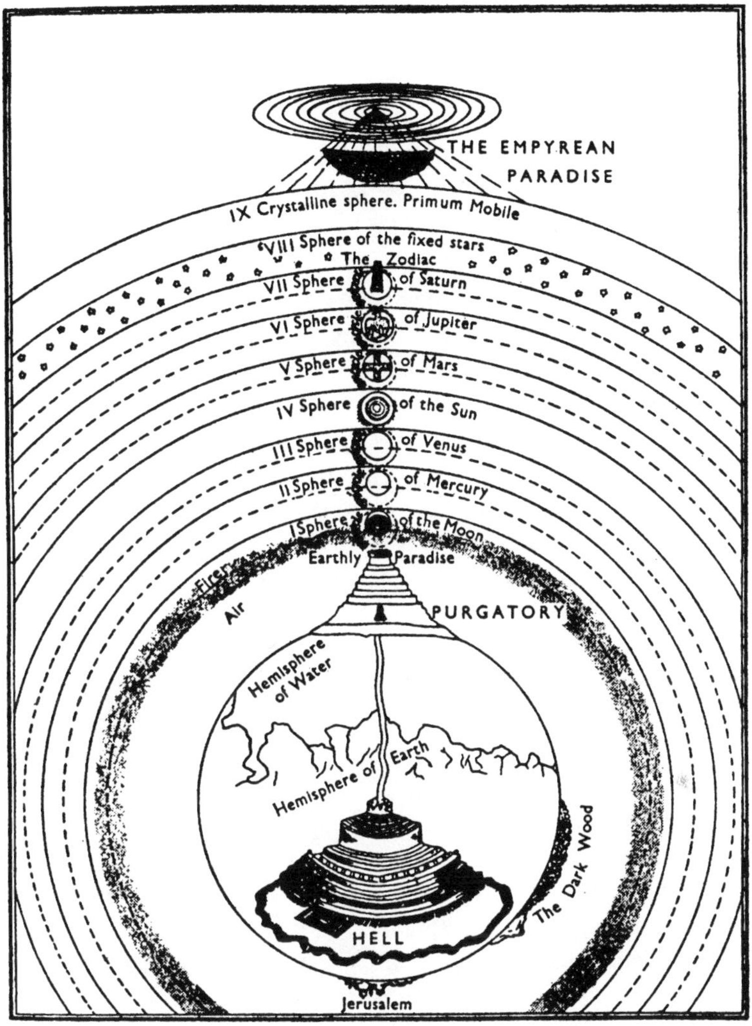

Dante's scheme of the universe

Popular accounts of modern science frequently paint Nicolas Copernicus (1473–1543) as a man who deliberately set out to overthrow this picture. They see him as having set science on a new road, which led straight through Galileo and Newton to our own times; and as fearlessly speaking truths which other men in his time were either too blind to see or too prejudiced to admit. His great book *On the Revolutions of the Heavenly Orbs*, on this view, is the first landmark in modern science, bringing about in astronomy a Copernican Revolution which quickly widened into a general Scientific Revolution. Once men had begun to trust the evidence of their own observations and experiments, and were no longer hidebound by tradition, Copernicus' discovery that the Earth went round the Sun was soon followed by other, equally revolutionary discoveries . . . or so the story goes.

We must now look past the half-truths of this caricature, to show what Copernicus attempted and what he in fact achieved. For in science, as in politics, the term 'revolution'—with its implication that a whole elaborate structure is torn down and reconstructed overnight—can be extremely misleading. In the development of science, as we shall see, thorough-going revolutions are just about out of the question. So much has to be done. New ideas are localized to begin with, as ways of dealing with certain specific problems. Having proved their worth as solutions of these particular difficulties, they must then meet wider objections within their original field of application; and their full possibilities and implications become apparent only gradually, as men in other fields build on the new constructions, and the rest of our ideas are reorganized to accommodate them. The profoundest changes in the long run may have had quite small beginnings.

Copernicus displaced the Earth from its position at the centre of the planetary system, in no wild spirit of rebellion against tradition. He did so simply because, in his opinion, Ptolemy's improvized constructions offended the intellect; and he wished to produce a more coherent system of planetary geometry. It took more than a century for his system to establish itself. For at least fifty years after his death, it was far from clear that its gains outweighed its losses, or that the objections to it could be met. Copernicus eventually succeeded where Aristarchos had failed, because he was fortunate in his successors. The theories of Kepler and Newton gave a new depth and significance to a change which, for Copernicus himself, represented little more than an intellectual prejudice. His quasi-aesthetic distaste for Ptolemy's mathematical methods provided his successors with a clue towards a new union of mathematical astronomy with astrophysics. Yet Copernicus himself was far from foreseeing the use that would later be made of his ideas: he saw himself, rather, as reinstating the claims of Aristotelian physics at the very points where Ptolemy had departed from it.

Copernicus was, in any case, not the first man in Western Europe to raise the question of the Earth's motion. By the early fourteenth century, all the important doctrines of antiquity were known, at any rate in outline, including those of Herakleides and Aristarchos. Nicolas of Oresme, a leading Parisian scholar, discussed at length the suggestion that the rotation of the Heavens might be an optical illusion, resulting from the daily rotation of the Earth. A century later, Cardinal Nicolas of Cusa also discussed the motion of the Earth, raising difficulties about the concepts of space and motion which are still unresolved today. So, whatever may be said about the poets and professional theologians of the Middle Ages, the traditional picture of the Heavens was certainly not accepted uncritically by all the learned doctors of the Church.

MEDIAEVAL ARGUMENTS ABOUT THE MOVING EARTH

Nicolas Oresme cut the ground from under the feet of Aristotle and Ptolemy by a series of arguments, many of which Copernicus himself happily took over. Oresme himself did not, in the end, accept the rotation of the Earth: his purpose was only to demonstrate that one could not dismiss the theory on grounds either of argument or of observation. His own view was that the question whether the Earth moved or was still must be judged by the same standards as articles of faith in religion—it was, in fact, a matter for 'natural reason' and revelation:

> Yet, nevertheless, everyone holds and I believe, that the Heavens and not the Earth are so moved, for 'God created the orb of the Earth, which will not be moved' (Psalm 92:1), notwithstanding the arguments to the contrary. For these arguments are persuasive, but do not make the conclusions obviously true. Having reflected on everything which has been said, one might accordingly believe that the Earth, rather than the Heavens, is moved in this way, and there is no evidence to refute this. Still, this conclusion

seems quite as much, or more, against natural reason as many Articles of our Faith.

To make his point, Oresme refuted in turn *all* the standard arguments against the idea of a rotating Earth. He tackled first the arguments from observation. If it is objected that the Heavens can be *seen* to move, then we can simply reply that all motion is relative:

Suppose that a man were located in the Heavens, so that they both moved together with a daily rotation; and suppose that this man, carried in the Heavens, had a clear and distinct view of the Earth, together with its mountains, valleys, rivers, towns and castles. It would then seem to him exactly as though the Earth were moving with a daily rotation, just as it appears to us on the Earth that the Heavens do. In the same way, if the Earth rotated daily rather than the Heavens, it would appear to us that the Earth was at rest and the Heavens moving. One only needs a little intelligence in order to imagine this easily. Thus the reply to this first observation is clear: one could perfectly well say that the reason why the Sun and the stars *appear* to rise and set as they do, and the Heavens *appear* to rotate, is the movement of the Earth.

He next shows how to answer the standard objections from dynamics. Must the Earth's rotation inevitably give rise to violent winds? No, answers Oresme: if we suppose that the atmosphere shares the rotation, this will not be so. All local motions on the Earth—even of arrows, clouds, and other things not actually in contact with the ground—will take place just as if the Earth were still. Observations of this kind cannot determine conclusively whether the Earth is rotating or not. This remains true today, despite all the changes in science since Oresme's time.

The reply to the second observation is as follows: the daily motion affects not only the Earth, but along with it the water and the atmosphere. . . . Think of the air enclosed in a moving ship: to a person in the ship, this air would appear to be stationary. . . .

Suppose [likewise] that a man were on a ship which was moving swiftly in an Easterly direction, without his being conscious of the movement; if he drew his hand downwards in a straight line, relative to the ship's mast, the hand would from his point of view seem to move with a single rectilinear motion only. According to [Herakleides'] opinion, an arrow shot upwards seems to us in the same way to rise or fall in a straight line.

This last argument was later repeated by Galileo, who is often credited with having originated it.

What, then, about computational astronomy? There, Oresme replies, the only difference introduced by supposing the Earth to rotate is a linguistic one. (Here again he is entirely correct.)

As for the fifth argument, according to which, if the Heavens did not rotate from day to day, all astronomy would be untrustworthy: this I simply deny. It is obvious from our reply to the first observation that the alternative view [i.e. the Earth's rotation] would leave all aspects, conjunctions, oppositions, constellations, figures and influences of the Heavens exactly as they are. Tables of movements and all other astronomical books would be just as true on the alternative opinion as they already are, except that one would speak of the daily rotation as taking place 'apparently' in the Heavens, but 'actually' in the Earth. There is no phenomenon which is consistent with the one view but not with the other.

Leaving aside the direct assertion in Psalm 92, indirect arguments from the wording of the Scriptures are no more conclusive. The Bible refers in the story of Joshua to the Sun 'standing still in the sky', as a miraculous exception to the general rule, but the Bible is expressed in the familiar idioms of everyday language so as to be intelligible to ordinary men. 'For instance, it is written that God is "repentant" or "pacified", and other such things, which do not mean just what they seem to do.' (This argument, too, Galileo was to repeat 300 years later: the grave offence he gave by doing so indicates the traumatic effect on the Church of events in the intervening years.)

One can go even further, Oresme continues: there are positive arguments for the view that the Earth *does* rotate. The idea that the entire Heavens move through a complete circle every day is in some ways even more difficult to accept:

One is forced to assume that their speed is excessively great. Anyone who reflects on the great height or distance of the Heavens, their dimensions and the length of their daily circuit, will see this clearly: if such a rotation is completed in a single day, one cannot imagine or conceive how the speed of the Heavens could be so marvellously and exceedingly great. . . .

Again, if one assumes that the whole Heavens move with a daily rotation, and that in addition the eighth sphere has another motion, as astronomers suppose [to explain the precession of the equinoxes], one is then forced to assume a ninth sphere, which moves with the daily rotation only. However, let the Earth move as has been said, and the eighth sphere need have a single slow movement only [i.e. precession]: so, on this theory, one need not dream up or imagine a ninth sphere in Nature, invisible and without stars.

The classical objections to a rotating Earth had, accordingly, been cleared out of the way more than a century before Copernicus was born. In the intervening years, some even more drastic questions were asked by Nicolas of Cusa, a fifteenth-century German cardinal, in the course of a general attack on dogmatism in natural philosophy, called *On Learned Ignorance*. The difficulties he raised about the concepts of cosmology were taken up again at the end of the eighteenth century by Immanuel Kant, a philosopher with first-hand experience of astronomy, and they are still alive in the current cosmological disputes involving such men as Fred Hoyle and George Gamow.

Cusa argued that the traditional world-picture could not really be taken literally. The outermost sphere, which was supposed to receive its motion directly from God, was treated as the boundary not only of the whole material universe but also of Space itself. Yet how could Space have a boundary? If you stood just inside this boundary and fired an arrow towards it, what was supposed to happen: would the arrow bounce back, or disappear entirely, or what? The whole idea of a bounded Space landed one in paradox—any 'boundary' must divide one spatial region from another, so that it was nonsense to talk of Space-as-a-whole having a 'boundary'. The same sort of paradox arose over all talk about the 'beginning of Time'.

Nor could one think of the Universe-as-a-whole having a 'centre' either. By this 'centre' we must mean the point which is, in every direction, symmetrically placed with respect to the 'boundaries'. If references to the boundary are unintelligible, any selection of a centre must be arbitrary. It may be convenient for certain purposes to regard the Earth as the centre; but nothing can *oblige* us to do so. No part of the universe—Earth, Sun, or anything else—has any unique right to be called the 'centre'. Observed motions are all relative, and it is a matter of decision what point in the universe is to be selected as the central origin of reference. So the view that the Earth is moving is just as admissible as the view that it is at rest.

Still, neither Oresme nor Cusa was a professional astronomer. Their arguments served only to raise questions, not to settle them. Neither Ptolemy's geometrical calculus nor the mediaeval cosmology could be expected to wither away, until some equally comprehensive alternative had been established. Copernicus was the first man who met Ptolemy on his own ground, and showed how to retain the strong points of his planetary geometry, while avoiding the objectionable features of his theory.

COPERNICUS: HIS AIM AND HIS THEORY

In reconstructing Copernicus' train of thought, the first thing to emphasize is this: he did not attack Ptolemy's planetary theory *because* it was Earth-centred. He was driven by dissatisfactions of a purely *theoretical* kind. The devices to which Ptolemy had resorted were unacceptable to him, and he was determined to replace them by others which were intrinsically more reasonable. He was led, in doing so, to re-order the planetary system in a new perspective, centred around the Sun instead of the Earth. The motion of the Earth was a *consequence* of this change, not its main aim, and was forced on him as the only satisfactory way out of an obstinate intellectual quandary.

Where Copernicus criticizes earlier astronomers, two phrases recur: their theories are 'inconsistent and unsystematic', and they 'violate the principle of regularity'. The first objection needs no explanation: he is affronted by the blithe way in which Ptolemy, without apology, shifts from one kind of construction to another, according to the particular problem he is tackling.

It is as though, in his pictures, an artist were to bring together hands, feet, head and other limbs from quite different models, each part being admirably drawn in itself, but without any common relation to a single body: since they would in no way match one another, the result would be a monster rather than a man.

Mathematically, the traditional teaching had given, not a coherent system, but only an unrelated jumble of constructions. His own system, he explains, is *truly* systematic:

If the movements of the remaining planets are related to the Earth's own circulation round its orbit, and are calculated in proportion to the sizes of each planet's circle, not only do the phenomena work out correctly, but the order and sizes of all the heavenly bodies and spheres, and that of the heavens themselves, become so interdependent that no part can be moved from its place without throwing all the parts, and the whole universe, into confusion.

His second objection to the Ptolemaic theory went deeper. The standard geometrical methods ran counter to the fundamental principles of physics. While paying lip-service to Aristotle's ideal of uniform circular motion, Ptolemy had introduced—e.g. by his use of 'equants'—irregularities which he left unexplained. It was in the hope of remedying this second defect, and bringing mathematical astronomy back into line with the Aristotelian ideal of uniformity, that Copernicus embarked on the construction of his rival planetary scheme.

This aim he states very clearly in the opening paragraphs of the *Commentariolus*, the sketch that he circulated privately some years before the *De Revolutionibus*.

Our ancestors, I observe, assumed the existence of a large number of celestial spheres for one reason especially: in order to explain the apparent movements of the planets in a way consistent with the principle of regularity. For they considered it absolutely absurd that a heavenly body, which is a perfect sphere, should not always move in a uniform manner. And they recognized that, by connecting and combining regular motion in various ways, they could make any body appear to move to any position. . . .

Yet the planetary theories of Ptolemy and most other astronomers, although consistent with the numerical data . . . present no small difficulty. For these theories were not adequate unless one also thought up certain equants: it then seemed that the planets moved with uniform velocity neither on their deferent circles nor around the centres of their epicycles. Such a system appears neither sufficiently absolute nor sufficiently attractive to the mind.

Being aware of these defects, I spent much time considering whether one might perhaps find a more reasonable arrangement of circles, from which every apparent inequality could be calculated, and in which every element would move uniformly about its own centre, as the rule of absolute motion requires.

Copernicus' fundamental aim was, therefore, to prove that the movements of the Sun, Moon, and planets formed a genuine system, whose elements were uniform, circular motions, related together in a consistent manner. Astronomy must return from Alexandria to Athens, and abandon the makeshift techniques introduced by Ptolemy. Looked at in detail, this might seem a retrograde step: since Newton's time, the paradigm of uniform circular motion has been completely abandoned. But *in spirit* the programme represents a real advance. Copernicus' insistence that the constructions of planetary geometry ought not to violate the accepted 'principles of regularity' paved the way for a reunion of astronomy with physics. From now on mathematical astronomy had once more to make sense in terms of the central ideas of physics.

So much for the task. If Copernicus had been able to improve on Ptolemy in every respect, the succeeding period of dissension—within science at any rate—might have been much briefer. But his reach exceeded his grasp, for reasons which, in retrospect, are clear enough. At first things went well. It became possible to account for the major planetary inequalities more naturally and simply than before. All that was required was seven initial assumptions:

1. There is no centre of all the celestial circles or spheres.

2. The centre of the Earth is not the centre of the Universe, but only of gravity [N.B. By this he meant 'weight', not 'gravitational force'] and the lunar sphere.

3. All the spheres rotate about the Sun as their midpoint, and so the Sun is the centre of the Universe.

4. The Earth's distance from the Sun is . . . imperceptible when compared with the height of the firmament [of fixed stars].

This fourth assumption is, of course, an attempt to forestall the standard objection based on the absence of stellar parallax: Copernicus simply echoes Aristarchos' own assertion about the remoteness of the stars—a view for which, incidentally, there was no independent evidence.

5. Any apparent motion of the firmament is the result, not of the firmament itself moving, but of the Earth's motion. The Earth, together with the material elements lying around it, goes through a complete rotation on its axis each day, while the firmament and highest heaven remain unaltered.

6. What appear to us as [annual] motions of the Sun result; not from its moving itself, but from the [linear] motion of the Earth and its sphere, with which we travel around the Sun just like any other planet. The Earth has, accordingly, more than one motion.

The Earth, then, not only rotates but travels in a complete orbit around the Sun; and in due course Copernicus gives it a third motion, to explain the precession of the equinoxes.

The final assertion expresses his crucial insight: namely, that many of the irregularities hitherto attributed to the planets were optical illusions.

7. The apparent retrogradations and [returns to] direct motions of the planets are the result, not of their own motion, but of the Earth's. The motion of the Earth alone, therefore, is enough to explain many apparent anomalies in the heavens.

At this first stage, Copernicus' new ideas served him very well. The results convinced him that his new picture was fundamentally correct. Ptolemy had introduced into every one of his planetary constructions a component having a period of precisely one terrestrial year. From the mathematical point of view, this might be a simple coincidence, yet a physicist would want to explain it. Copernicus accounted for all these annual components as common by-products of the Earth's own motion. In this way, he converted the seemingly complex sequence of retrograde loops—familiar and puzzling since the earliest times—into something conforming closely to Aristotle's ideal. Further, by comparing the sizes of the loops, he was at last able to estimate the relative sizes of the planetary orbits reliably and consistently. Finally, Copernicus began to build up a detailed and coherent planetary system, using as elements only uniform circular movements. Others had attempted to do this before him: at least one Arab astronomer had shown that equants could be dispensed with if one used a large train of epicycles instead. By re-ordering the whole system around the Sun instead of the Earth, Copernicus became the first man to carry the programme through to completion. (Contrast Plates 9 and 10.)

Up to this point all had gone well. There were, of course, difficulties: some of them astronomical, others dynamical. These he tried to meet in his final account, *On the Revolutions*, published in the year of his death. The absence of stellar parallax compelled him, as it had Aristarchos, to think of the fixed stars as at a distance in comparison with which the diameter of the Earth's annual orbit shrank to nothing. Then, there was a problem about the planet Venus. If its greatest distance from the Earth was over four times its least, one would expect its brightness to vary much more strikingly than in fact occurred. (The phases of Venus provided the answer to this question, and were discovered later by Galileo. Venus and the Earth are furthest apart when they are on opposite sides of the Sun. Just before disappearing behind the Sun, Venus displays its full disc to the Earth, so counteracting the effects of distance: at its nearest, when you might expect it to be very bright, it shows only a thin arc, like the New Moon.)

Lastly, there were the dynamical difficulties, and Copernicus adapted the arguments of Oresme:

> Why then do we still hesitate to allow the Earth the mobility naturally appropriate to its spherical shape, instead of proposing that the whole Universe, whose boundaries are unknown and unknowable [cf. Cusa], is in rotation? And why do we not grant that the daily rotation of the heavens is only apparent, while that of the Earth is real? It is like what Aeneas said in Virgil's Aeneid (III.72): 'We sail out of the harbour, and the land and cities retire.' When a ship floats along on a calm sea, all external things appear to the sailors to be affected by a motion which is really the motion of the ship, while they themselves seem to be at rest along with everything which is

with them on the ship. Doubtless, in the case of the motion of the Earth, it could happen similarly that the whole Universe was thought to rotate. . . .

As to [terrestrial] things which rise and fall, we must allow that their motion relative to the whole Universe is a double one, being generally the combination of a rectilinear motion and circular one. Things whose earthiness predominates are borne downwards by their weight . . . things which are fiery are carried up into the higher regions.

Only objects which are away from their natural places have this double motion: once earthy things reach the ground or fiery ones the upper atmosphere, the rectilinear component ceases. Everything in its natural place moves in a simple circle, for

a circular motion is always uniform, since it has a never-failing cause of motion.

So far Copernicus has been establishing his new astronomy by relying on the principles of Aristotelian physics. But now he gives Aristotle's ideas a new twist. If we are to choose between a daily rotation of the Earth and a similar rotation of the Heavens, he argues, surely 'the immovable sphere of the fixed stars, which contains and gives position to all things', should be considered at rest—rather than the changing and unstable Earth. In the case of the annual cycle of the seasons, he argues similarly: the corrupt and changeable Earth has no claim on the central place, when compared with that splendid source of light and heat—the Sun. A religious respect for the Sun had been for many centuries a feature of Babylonian and Persian religion, and had survived into Christian thought through the influence of the Alexandrian neo-Platonists: it now plays a part in Copernicus' thought.

In the middle of everything stands the Sun. For in this most beautiful temple [the Universe], who could place this Lamp [the Sun] in any other better place than one from which it can illuminate all other things at the same time? This Sun some people appropriately call the Light of the World, others its Soul or its Ruler. Trismegistus calls it the Visible God, Sophocles' Electra calls it the All-Seeing. Thus the Sun, sitting on its Royal Throne, guides the revolving family of the stars [i.e. the planets].

Copernicus must have found his theory very attractive. It was systematic, consistent, and coherent, as Ptolemy's system had never been; it was free of the offences against the 'principle of regularity' introduced by the use of equants; and it restored to the Source of Light and Heat the rightful place allotted by the Pythagoreans to the Central Fire. His obstinate search for a more rational account of the planetary system seemed justified.

Unfortunately, matters did not go on as well as they had begun. If he was to convince his fellow-astronomers that his own system was superior, it was essential to equal Ptolemy in the scope of his calculations. That meant, in effect, re-writing the *Almagest* completely. The greater part of his book *On the Revolutions* embodies the result. However, by the time that Copernicus had finished, there were as many complexities of detail in his system as there had been in Ptolemy's. In the *Commentariolus* he had promised to make do with only 34 spheres: to fit the facts, he had to introduce a good dozen more. Although he refused to admit equants, he could not get along without eccentrics and epicycles; and, when he really got down to the geometry, the Sun could not even remain at the exact centre of any of the planetary circles. He continued to keep the Sun stationary at the centre of the fixed stars, but was forced to refer the circles of all the planets to a point in empty space. This point was itself carried periodically around the Sun along an epicyclic track, and in turn was the moving centre of the Earth's motion. Even after all these complications had been introduced, the resulting calculation were little improvement in accuracy on Ptolemy's—sometimes better, sometimes worse.

The final outcome of Copernicus' work was, thus, something of an anticlimax. Bearing this in mind we may understand rather better why it was not more enthusiastically received. Unless one shared his passionate conviction that 'regularity' was worth saving at all costs, the intellectual price of his theory seemed hardly worth paying. Before 1600, there was only a handful of convinced Copernicans. Most of his successors reacted in a frankly pragmatic

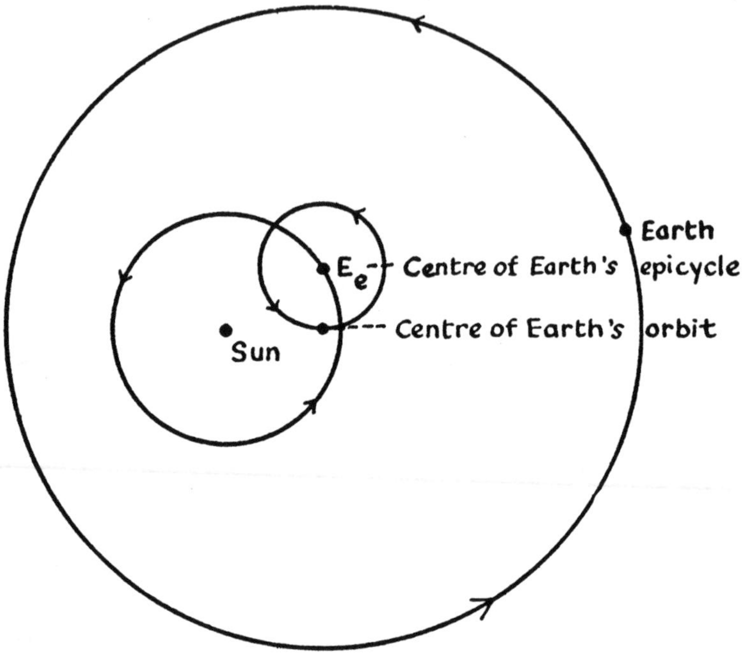

The motion of the earth on the Copernican system

The Earth revolves in a circle whose centre is not the Sun but a point eccentric to it. The centre of the Earth's orbit is itself a moving point, moving round E_e which is moving in its turn round the Sun

way: they ignored his principles, but took over his methods of calculation wherever they proved more accurate than Ptolemy's. For other problems, they had no hesitation in hanging on to the old methods: the inconsistency of treating the Sun sometimes as moving, sometimes as stationary, they ignored. Fifty years later, the English astronomer Thomas Blundeville still assumed that the Earth was stationary, but added:

> Copernicus . . . affirmeth that the Earth turneth about and that the Sun standeth still in the midst of the Heavens, by help of which false supposition he hath made truer demonstrations of the motions and revolutions of the celestial spheres, than ever were made before.

One thing about the published version of Copernicus' book was designed to encourage this attitude. For with it there appeared an unsigned preface by his colleague Osiander. First reports of Copernicus' new system, as Osiander knew, had troubled Protestants and Catholics alike. The traditional symbolism of the mediaeval world-picture was so much a part of everyone's ideas that Copernicus' innovation could not be confined to astronomy: it inevitably had wider repercussions. So he prudently added a few words addressed to the Reader, 'Concerning the Hypotheses of this Work'. There was no need to take offence, he explained, since

> the author of this work has done nothing blameworthy. For it is the duty of an astronomer to compose the history of the celestial motions through careful and skillful observation. Then . . . he must conceive and devise, since he cannot in any way attain to the true causes, such hypotheses as, being assumed, enable the motions to be calculated correctly from the principles of geometry, for the future as well as for the past. The present author has performed both these duties excellently.

No one, Osiander says, could suppose that Copernicus was describing the *true* structure of the planetary system, and he mentions the difficulties about Venus. But then, he goes on to argue—shifting his ground from astronomy to metaphysics—it is not the business of astronomers to do that:

> It is quite clear that the causes of the apparent unequal motions are completely and simply unknown to this art. And if any causes are devised by the imagination, as indeed very many are, they are not put forward to convince anyone that they are true, but *merely to provide a correct basis for calculation.* Now when from time to time there are offered for one and the same motion different hypotheses . . . the astronomer will accept above all others the one which is the easiest to grasp. The philosopher will perhaps seek the semblance of the truth. But neither of them will understand or state anything certain, unless it has been divinely revealed to him.

Having given the theologians the last word in cosmology, Osiander invites astronomers to treat Copernicus' system exactly as, in the event, they did:

Let us therefore permit these new hypotheses to become known together with the ancient hypotheses, which are no more probable; let us do so especially because the new hypotheses are admirable and also simple, and bring with them a huge treasure of very skilful observations. So far as hypotheses are concerned, let no one expect anything certain from astronomy, which cannot furnish it, lest he accepts as the truth ideas conceived for another purpose, and depart from this study a greater fool than when he entered it. Farewell.

The author of these words is Copernicus' friend, Osiander; but the sentiments—which are in contradiction to Copernicus' own explicit statements elsewhere—are those of Ptolemy and Oresme. Copernicus had not asserted the motion of the Earth 'gratuitously' or 'merely to provide a basis for calculation'. He had produced strong arguments for this motion, based on his conviction that celestial changes happen uniformly. He had admitted in the *Commentariolus* that 'the planetary theories of Ptolemy' were 'consistent with the numerical data': it was their *irrationality* that he objected to. This was his complaint, and this was what he set out to remedy.

Contrary to Osiander's interpretation, Copernicus believed that, despite everything, the truth about the Heavens *could* be discovered by rational investigation. This meant claiming back from the theologians the right to pronounce on cosmological questions, and bringing cosmology back from the sphere of revelation into the scope of reason. In the long run, a clash with the religious authorities was probably inevitable. For the time being, Osiander did his best to prevent it, and for fifty years the implications of the Copernican view were not generally recognized. At the turn of the century, Galileo and Kepler—among others—began to insist publicly on what Osiander had tried to hush up. In the meanwhile, Copernicus' views had very nearly gone the same way as those of Aristarchos.

If they had done so, this would not have been entirely surprising. In retrospect, we are tempted to pay too much attention to the things that happened after Copernicus, and so fail to see his theory as a contemporary would have seen it. After all, he made no new striking discoveries about the Heavens, such as Tycho was to make after him. Indeed, he recorded only a few dozen astronomical measurements—just so many as he needed, to tie his own calculations in with the far more voluminous records left by Ptolemy. Generally, he was content to accept as reliable the much-copied and partly-corrupt Ptolemaic records. His methods of astronomical computation were not, in principle, any more accurate than Ptolemy's: in practice they were sometimes less so. Nor were they, in the end, appreciably simpler than the earlier methods. Even the outstanding merit of the system—the recognition that retrograde motion was an optical illusion—was obscured by the time the theory was completely worked out; in the end, the constructions used to compute the planetary motions reintroduced elements from the Earth's own motion, by way of the 'moving centre' of the Earth's circular orbit.

In the face of all these negative results, why is Copernicus thought of as initiating a revolution? The answer is: his new system was put forward in the right place at the right time. In almost every respect, Copernicus was an ancient astronomer, not a modern one. His ideas have more in common with Ptolemy's and Aristotle's than they had with Kepler's or Newton's. (At first glance, it is not easy to tell a page of Copernicus' *On the Revolutions* from a page of Ptolemy's *Almagest.*) But, in the seventy-five years following Copernicus, evidence and arguments began to accumulate which, taken together, gradually eroded the classical picture of the cosmic system, and—more important—suggested the outlines of a possible alternative.

Most important of all: Copernicus had started astronomy on the road back to physics—though, paradoxically, rationality was exemplified for him by a paradigm we now reject: circular motion. As a physicist, Copernicus introduced only one important innovation. In explaining away the apparent daily motion of the sphere of fixed stars, he removed from Aristotle's system the 'unmoved mover': that self-moving sphere which, through its connections to the planetary spheres within it, controlled and regulated the movements of the whole. This change made it necessary to look in a new direction for this control, so clearing the ground for novel theories of celestial force: e.g. Kepler's idea that the planetary motions are controlled, not from the outside, but from the inside—by the Sun.

Standing at the very beginning of modern physics, Copernicus had—and could have—no clear vision of what he had started. Without his new heliocentric perspective, no adequate system of celestial physics could have been built up. But his geometrical re-ordering of the planetary system was only a beginning. Half a dozen other steps were required, both in astronomy and in dynamics, before the intellectual revolution could be completely carried through. The final result would have amazed even Copernicus.

FURTHER READING AND REFERENCES

Little in the way of general material exists in English on science in the Byzantine and Islamic worlds. The best introduction to Arabic science (though one more useful to scholars than to the general reader) remains

A. Mieli: *La Science Arabe*

There is a good general account of life and thought in Byzantium in

S. Runciman: *Byzantine Civilization*

but this refers only fleetingly to scientific ideas.

There is an admirable account of the reworking of the classical tradition in cosmology to form the mediaeval world-picture in

T. S. Kuhn: *The Copernican Revolution*

For the general development of science in the middle ages, see

A. C. Crombie: *Mediaeval and Early Modern Science* (first published as *Augustine to Galileo*)

Charles Singer: *From Magic to Science*

The intellectual and social background to the rise of Mediaeval Europe are discussed admirably in

R. W. Southern: *The Making of the Middle Ages*
H. Pirenne: *Economic and Social History of the Middle Ages*

For the arguments of Oresme, see the detailed discussion in

M. Clagett: *The Science of Mechanics in the Middle Ages*

For the work of Copernicus, the indispensable book is

E. Rosen: *Three Copernican Treatises* (second edition, with additions, recently issued by Dover Books)

Useful material about Copernicus can be found in Kuhn's book, as well as in

A. Armitage: *The World of Copernicus* (also published under the title *Sun, Stand Thou Still*)

H. Butterfield: *The Origins of Modern Science*

The opening section of Part II of the film *Earth and Sky* deals with the transition to the Copernican system, demonstrating the geometrical equivalence of the Ptolemaic and Copernican constructions, and showing how the Copernican perspective explained the retrograde motion of the planets away as an optical illusion.

7

Preparing the Ground

If Copernicus was in many ways a 'conservative', and even a 'reactionary', the changes in science which took place in the 150 years after his death were 'revolutionary'. But these historical labels again need careful qualification, and can be misleading. In two respects, certainly, these years saw a complete transformation in science: in the external conditions of scientific work, and in men's basic framework of cosmological ideas. Still it would be a mistake to suppose that, in the years around 1600, the whole of science—or even the whole of physics—was overturned and refurbished. In some directions, the period saw only an acceleration of changes already well under way during the late mediaeval period. In other sciences, so far from there being a revolution, men's state of understanding in 1700 was very much what it had been a couple of centuries before.

The progress made between 1550 and 1700—though very striking—was localized. The subjects in which the most dramatic changes can be found are those we are concerned with in the present book: astronomy and dynamics. Here the period ends with the establishment of Newton's new and comprehensive system of the world. For the moment, we must look at the steps by which the ground was prepared for this new system of thought.

THE BACKGROUND OF THE NEW SCIENCE

There was at this time one profound change in men's attitude to science which, though intangible, was of the highest significance. Throughout the seventeenth century there was a rapidly growing number of men in Western Europe who felt the same confidence as the early Greek philosophers in the power of the intellect, and in the potentialities of mathematical reasoning allied with honest observation. Even theologically men were now prepared to justify the free use of one's eyes and one's mind in the interpretation of natural happenings. God—it was now said—gave Man two separate sources of evidence about His Creation: the Book of Holy Scripture, and 'the Book of Nature': He was not only the Creator, but also the Divine Architect, whose Mind could be read by studying the products of His workmanship. So it ceased to be the mark of improper curiosity to engage in scientific enquiries: they became rather a pious duty. Where Aquinas had earlier done his best to harmonize the teachings of Aristotle with the doctrines of the Church, Protestant theologians by 1650 were trying to do the same for the new ideas and discoveries of science.

The revival of confidence was given a great fillip by one technological innovation—one hesitates to call it an invention, since the Chinese had practised the art for centuries—namely, printing. All over Europe, from 1500 on, there was a great burst of intellectual activity, as the printing-press multiplied the number and variety of books in general circulation. Men could now study at first hand far more texts than they could ever hitherto have afforded. It became economic—as in our present 'paper-back revolution'—to produce new and better editions of the classics: the re-publication in Venice of Archimedes' original works stimulated a new wave of mathematical research, and from now on the learned men of Europe could get their works into the hands of their colleagues in many countries in the course of a few months.

It would be hard to exaggerate this change in the efficiency of scientific communication. Whereas the pace of scientific advance had depended earlier largely on the concentration of scholars in one and the same place, from now on men living at a distance from one another could collaborate effectively. Before 1500, one can hardly speak of mathematical or scientific advances ever becoming 'common knowledge': important insights were achieved at one place and time, only to be lost again in the next century, and scholars working in different cities took as their starting-points quite different bodies of knowledge. From about 1550 one can fairly assume that anyone writing about, say, planetary dynamics has read most of the existing material directly relevant to his work. Only in the twentieth century, when the stream of scientific publications has turned into a flood, does this cease to be a reasonable assumption; the age of original papers and books is being rapidly replaced by the age of 'abstracts'.

One last factor of real importance must be mentioned—the rise of national scientific academies. Public men such as Francis Bacon, Earl of Verulam (1561–1626), began to preach the benefits which would come to mankind from science, and called for the establishment of institutions to encourage scientific research. In Italy, France, and Britain, ruling families were persuaded to give to learned academies their patronage, and in some cases also their financial support. These academies, notably the Accademia Dei Lincei, to which Galileo belonged, and the Royal Society of London, founded in 1660, were the active centres of scientific discussion and publication.

THE WORK OF TYCHO BRAHE

Copernicus' new astronomical system left much of Aristotle's synthesis untouched, and its merits were paid for at a high price. His most notable immediate successor, Tycho Brahe (1546–1601), introduced his own drastic amendments to the classical picture, but declined to follow Copernicus all the way. Tycho, rather than being an interpreter of Nature, was above all a passionate observer of the Heavens. He embarked on a more systematic and continuous programme of astronomical observations than had been made by anyone hitherto. The traditional practice was to record the positions of the planets, only at significant points in their tracks—for instance, stationary points in the retrogradation. Tycho saw that a precise determination of the shapes of the orbits—or indeed any well-established planetary theory—must be based in the long run on more comprehensive and accurate planetary records than the tradition preserved. At his private observatory in Denmark, and for a short time at Prague, where he became chief mathematician and astronomer to the Emperor, he recorded the positions of the planets, notably of the planet Mars, night after night. (Plate 11.)

As he was an observer, it is perhaps not surprising that his crucial objection to the Copernican system was based on a simple observation: namely, the absence of stellar parallax. The patterns of the constellations remain unchanged from season to season, while (if Copernicus was to be believed) the Earth travels more than a hundred million miles. This motion must inevitably produce some visible change in the constellations unless, by comparison with the distance of even the very nearest stars, the diameter of the Earth's orbit is little more than a geometrical point. Yet most of the traditional pictures of the Heavens showed the fixed stars only a little further away than the planet Saturn, so that banishing them to an enormously greater distance required an immense effort of the imagination. The more Copernicus' successors looked at this difficulty, the more serious it began to appear. Even Kepler, who was a confirmed Copernican from his student days, spent a considerable time trying to observe the parallax. Had there been a single star whose parallax could be detected with the naked eye, the triumph of the heliocentric perspective would have been much swifter. (Plate 6.)

It was not just a matter of *distances*. Further difficulties arose over the size and distribution of the fixed stars also. If, as Copernicus argued, the nearest stars were so distant that a terrestrial observer could move 200 million miles without their patterns altering, why did the individual stars look so *large*? At that distance, one would hardly expect to see them. If any large star had shifted its position through a distance larger than its own apparent width, then, by careful measurement, Tycho could have detected the fact. As no such change was actually detected, it seemed to follow from the Copernican theory that the nearest and brightest stars must all of them be several hundred million miles in diameter—in order to appear to us as they do. In that case, the Sun would be in size only a small fraction of any one of the fixed stars. This made the price of Copernicus' innovations even greater than at first sight appeared.

Tycho could not reconcile these three things: the motion of the Earth, the distance of the stars, and the belief that the Sun is as large as an average star. And, as things then stood, he had every reason to believe that those things were incompatible. There was in fact an unexpected snag, which did not become apparent until after Galileo's work with the telescope. For any estimate of the real size of the stars based on their apparent diameters assumed that the spots of light we see in the sky are a reliable guide. As we know now, starlight is blurred by diffraction as it passes through the pupil of the eye, in just the same way as the light from a distant torch or lamp. One cannot estimate the size of a distant star from its *apparent* diameter any more than one can the size of a distant light seen from miles away across country.

So, though Tycho was prepared to recognize all the clear advantages of the Copernican system, on one point he was adamant. He conceded that the five regular planets all travelled in circles around the Sun; yet, as he pointed out, it did not at all follow that one had to set the Earth in motion as well. There was a third possibility: a system neither Ptolemaic nor Copernican, but one which—by keeping the Earth at rest—preserved the chief intellectual merits of Copernicus' system without incurring its difficulties. To account for retrograde motion, we need suppose only that the five major planets went around the Sun: the Sun and its five satellites could then all move together around the Earth. The Copernican calculations would remain the same, and no astronomical difficulties would arise. Two such systems were in fact possible. In Tycho's own system, the starry sphere rotated around a central, stationary Earth, while the Sun and planets lagged behind the stars by an average of one degree a day. In the alternative system put forward by von Baer, the stars were still and the Earth spun once a day, while the Sun went round it once a year. In either case, the distance from the Earth to the fixed stars remained unaltered, as in the Ptolemaic system. (Plate 12.)

The resulting 'Tychonic' system had great attractions. For those who liked to think of themselves as 'conservative moderns' it provided a convenient middle way. John Donne, for instance, was heavily sarcastic about Copernicus in *Ignatius His Conclave*, and was inclined to favour Tycho's compromise. Its attractions are clear enough. So long as the sphere of fixed stars was thought of as enclosing and defining the positions of things, the parallax problem was a real stumbling-block: it appeared rather gratuitous to suppose that the Earth was in motion round the Sun when—as Tycho pointed out—there really was no necessity to do so.

If Tycho had contributed nothing more to astronomy than his compromise system, he would not be so significant a figure. In fact, he provided also two new pieces of evidence that the traditional picture of the Heavens would have to be altered. The initial observations were made almost incidentally, in the course of his regular studies of the planets. But Tycho at once saw their importance. They are of great interest for two reasons. Quite apart from their impact on men's traditional beliefs, they show also, in retrospect, how far an observer tends to notice only those things by which he is prepared to be surprised. Events such as Tycho observed in the sky during the 1570's had occurred before and have happened since, without creating the same excitement. Indeed, they must have happened during the mediaeval period, when the traditional picture was firmly accepted, but their significance was at that time overlooked, for lack of any observers ready to recognize their importance.

The first observation was related to the immutability of the Heavens. In the year 1572, a bright light appeared in the neighbourhood of Cassiopea, and grew rapidly in intensity until it was as bright as the brightest stars in the sky. It appears to have been what is called nowadays a 'super-nova': a star which had previously been faint suddenly blowing up and gradually burning out. It was a piece of great luck for Tycho that this particular nova appeared when it did. Novas of this size appear on the average only once in three centuries. Observers all over Europe were excited by the appearance of the star and dozens of pamphlets were published about its implications. It could not be an atmospheric phenomenon, or its position in the sky would have varied from place to place on the Earth. As the years went on, people began to watch with growing curiosity to see whether it would move in the way a comet would do, or whether it was stationary and so presumably on 'the sphere of fixed stars'. As the finest observer of his time, Tycho was able to establish that, despite all rumours to the contrary, the light was stationary and in all respects indistinguishable from a normal first-magnitude star.

By 1570, men were ready to believe their eyes when a new star entered the Heavens. But it is certain, looking back, that a similar super-nova had been visible in Western Europe in the year A.D. 1054. A slowly fading star in the constellation of the Crab is believed to be the present-day stage of this super-nova, and the Chinese recorded its original appearance. Yet there is no record at all that anybody in Western Europe so much as noticed it. Of course, at that stage people's interest in the sky was almost entirely practical, and Aristotle's cosmology was not yet generally known in Europe; still, it is a small, though significant, indication of the scale of scientific activity in Europe in the eleventh century that so striking and anomalous a celestial occurrence should have gone completely unremarked. Perhaps, after all, the nova in the Crab did catch the eye of some European monk, and momentarily surprised him. But with only fragmentary star-catalogues and verbal tradition to go on, he would have hesitated before accepting his observation as authentic evidence of superlunary change. He would not have believed his eyes.

Tycho's second crucial observation was made five years later, in 1577. In that year a great comet appeared in the sky and remained visible for some months. Tycho followed its movement carefully, comparing its changing position with that of the stars, the Moon, and other heavenly objects. From all the observations he collected, he became convinced that the comet was not a sublunary object—for example, an atmospheric phenomenon like the Northern Lights—as the common doctrine implied. Rather, it was moving along a track far beyond the Moon. He confirmed this trigonometrically.

The star in Cassiopea was undoubtedly a new one, so it could no longer be held that the sphere of the fixed stars was utterly and completely free from change. The comet of 1577 threw doubt on yet another point in the traditional cosmology: the existence of solid planetary spheres. The comet was evidently moving, in a smooth and unhindered manner, right through the region between the planets in which these spheres were supposed to be located. (Plate 7.) By itself the observation was not conclusive. Nothing decisive followed from it, since it was not essential to think of the planetary spheres as solid, rigid, and complete. But Tycho's observation naturally led astronomers to ask, once more, just how securely based were the chief features of their cosmological picture. So long as theory left no place for comets in the superlunary regions, men decided, naturally enough, that they *must* be a feature of the sublunary world. But now this option no longer remained. Admitting comets into the superlunary world at once reopened the whole question of the *mechanism* of the planetary system, and this time it remained open for over a century, until it was settled by Newton's theory of gravitation.

Tycho had by now carried the criticism of the old cosmology in a direction which Copernicus never foresaw. But even Tycho, while he questioned the immutability of the Heavens and the existence of the crystalline spheres, still accepted the spherical shell carrying the fixed stars as the outer boundary of the universe, placing it at no immense distance from the Earth. Tycho was probably one of the best instrument-makers ever to work in the field of astronomy: his quadrants and other direction-finders were made with great accuracy and precision. He was also a scrupulous, honest, and systematic observer. Whatever men could have discovered while restricted to the naked eye, Tycho was capable of discovering. His limitations were entirely those imposed on him by this restriction. The next step in the erosion of the mediaeval world-picture came some thirty years later, as a result of Galileo's work with his telescope: the effect of this step was, among other things, to remove from our picture of the universe the traditional boundary which Tycho had still retained.

GALILEO'S TELESCOPIC DISCOVERIES

Galileo Galilei (1564–1642) is still a controversial figure today, as he was in his own lifetime. Forthright, open-minded, ingenious and inventive, he found it hard to be diplomatic or to conceal his opinions: he preferred to throw himself into polemical debates where he could exercise his brilliant talents, and the ease with which he so often discomfited his opponents won him not only devoted friends, but also influential and jealous enemies.

A list of Galileo's excursions into science includes almost every topic under serious discussion around 1600—at any rate in the physical sciences. At one time he was inventing a thermometer, in the hope of replacing subjective estimates of warmth and cold by objective numerical measurements. At another, he was concerned with problems of military engineering or the breaking-points of beams. Acoustics, hydrostatics, the vacuum, light, magnetism: he turned his attention to each subject in turn. And throughout his career, from his early days at Pisa, through the years of maturity at Padua and Venice, to his sad old age in Florence, two subjects preoccupied him: Copernican astronomy, and the mathematical theory of motion.

In both these fields he made important contributions, which later in the century were to be integrated into Newton's theory of planetary dynamics. Still, Newton was born only in the year that Galileo died; and it is important not to read back into Galileo later results which he did not foresee. In fact, the problems of astronomy and the problems of dynamics were, for Galileo, separate problems. In astronomy, he helped to break down the remaining barriers between the sublunary and the superlunary worlds. In mechanics he established the geometry of terrestrial motion on a new basis. But the final task, of extending his own mechanical discoveries into the Heavens and so accounting for the interactions of the Sun and planets, was one he seems never to have envisaged. Like Copernicus, he continued to accept unending circular motion as entirely self-explanatory; and—as we shall see later—his mechanical experiments even appeared to confirm this view.

Galileo made only one serious excursion into astronomy, and that on a limited front; yet its direction was well-chosen, and he exploited his discoveries to the hilt. The appearance of the new star of 1572 had brought into question the whole status of the fixed stars; and, as early as 1576, the Englishman Thomas Digges, in a popular account of the Copernican theory, had depicted them as stretching without limit in every direction around the central empty region containing the solar system. This new vision of an infinite and unbounded cosmos had been enthusiastically preached by Giordano Bruno, who had gone on to argue—as Copernicus had never dreamt of doing—that there was an infinity of worlds, containing many solar systems, some of them populated by human beings like ourselves. Even by the standards of twentieth-century science fiction, Bruno's cosmology was speculative, and he was executed as a heretic in the year 1600.

Bruno's abolition of the traditional *habitaculum Dei* and mansions of the elect, which even Digges had retained, provoked at last hostile reactions from the religious authorities. From this time on, the Copernican doctrine was officially under suspicion, and free thought in astronomy began to be distrusted in a way it had not been before. Yet ideas cannot be burned at the stake as heretics can. The idea that the stars went on indefinitely in every direction, and were not all equidistant from a single centre, was soon regarded as a serious possibility. So, by the time that Galileo turned his newly built telescope on to the Heavens, men's minds were prepared to receive a radically new account.

Galileo gave this account in his little book the *Sidereus Nuncius*, or *Starry Messenger*, written in Venice in the year 1610. He began by describing the telescope he had made, whose principle had been suggested to him by reports of similar instruments built in Holland for terrestrial use. Galileo's telescope multiplied thirty times in diameter and, though its discrimination was naturally not of the highest, the things he observed through it were striking enough. As he says:

> The number of the Fixed Stars which observers have been able to see without artificial powers of sight up to this day can be counted. It is therefore decidedly a great feat to add to their number, and to set distinctly before the eyes other stars in myriads, which have never been seen before, and which surpass the old, previously known, stars in number more than ten times.
>
> Again, it is a most beautiful and delightful sight to behold the body of the Moon, which is distant from us nearly sixty semi-diameters of the Earth, as near as if it was at a distance of only two of the same measures.

Characteristically, Galileo reports the results of his very first observation in terms which throw fresh doubt on the supposed perfection of the heavenly bodies. The whole face of the Moon, he reports, is covered with spots, smaller than those which had long been visible to the naked eye, so thickly scattered that they sprinkle the whole surface of the Moon, but especially the brighter portion of it. . . . From my observations of them, often repeated . . . I feel sure that the surface of the Moon is not perfectly smooth, free from inequalities and exactly spherical, as a large school of philosophers considers . . . but that, on the contrary, it is full of inequalities, uneven, full of hollows and protruberances, just like the surface of the Earth itself, which is varied everywhere by lofty mountains and deep valleys.

This opinion he backs up by describing the way in which the appearance of these spots changes from one part of the month to another.

They have the dark part towards the Sun's position, and on the side away from the Sun they have brighter boundaries, as if they were crowned with shining summits. Now we have an appearance quite similar on the Earth about sunrise, when we behold the valleys, not yet flooded with light, but the mountains surrounding them on the side opposite to the Sun already ablaze with the splendour of his beams; and just as the Sun rises higher, so also these spots on the Moon lose their blackness as the illuminated part grows larger and larger.

Next, he clears out of the way Tycho's difficulty about the apparent size of the stars. Galileo was at once struck by the fact that the stars he looked at were magnified very much less than the distance between them. The stars themselves remained points of light, while the stretches of black sky between them expanded many times. So the apparent diameter of the stars was, after all, misleading. He could not know that the visual discrimination one can achieve is limited, ultimately, by the aperture of one's instrument—whether this is the pupil of one's eye or a telescope: this became intelligible only in terms of the wave-theory of light. But at any rate his observations settled the chief point:

A telescope which (for the sake of illustration) is powerful enough to magnify other objects a hundred times, will scarcely render the stars magnified four or five times. . . . When stars are viewed with our natural eyesight they do not present themselves to us of their bare, real size, but beaming with a certain vividness and fringed with sparkling rays, especially when the night is far advanced; and from this circumstance they appear much larger than they would if they were stripped of those adventitious fringes. . . .

telescope . . . removes from the stars their adventitious and accidental splendours before it enlarges their true discs (if indeed they are of that shape), and so they seem less magnified than other objects.

(Even today in the twentieth century the most powerful telescopes can give us only the diffracted images of the stars: the planets alone are visible in their actual shape.)

The difference between the appearance of the planets and the fixed stars seems also deserving of notice. The planets present their discs perfectly round, just as if described with a pair of compasses, and appear as so many little moons, completely illuminated and of a globular shape; but the fixed stars do not look to the naked eye bounded by a circular circumference, but rather like blazes of light, shooting out beams on all sides and very sparkling, and with a telescope they appear with the same shape as they are viewed by simply looking at them.

His new 'telescopic' stars presented 'an infinite multitude'. Astronomers had previously classified the stars in six orders of magnitude, according to their visible brightness:

But beyond the stars of the sixth magnitude you will behold through the telescope a host of other stars, which escape the unassisted sight, so numerous as to be almost beyond belief, for you may see more than six other differences of magnitude, and the largest of these, which I may call stars of the seventh magnitude, or the first magnitude of invisible stars, appear with the aid of the telescope larger and brighter than stars of the second magnitude seen with the unassisted sight.

He had originally planned to depict the entire constellation of Orion, 'but I was overwhelmed by the vast quantity of stars and by the want of time', so he drew a picture of Orion's belt and sword, in which the three visible stars of the belt and the six in the sword were accompanied by 'eighty other stars recently discovered in their vicinity'. (Plate 6d.) The Pleiades similarly contained forty stars normally invisible, in addition to the six familiar ones.

He then turned to the Milky Way and enjoyed another cheerful slap at his academic colleagues.

By the aid of a telescope anyone may behold this in a manner which so distinctly appeals to the senses that all the disputes which have tormented philosophers through so many ages are exploded at once by the irrefragable evidence of our eyes, and we are freed from wordy disputes upon this subject, for the Galaxy is nothing else than a mass of innumerable stars planted together in clusters. Upon whatever part of it you direct the telescope straight away a vast crowd of stars presents itself to view; many of them are tolerably large and extremely bright, but the number of small ones is quite beyond determination. . . .

Further—and you will be more surprised at this—the stars which have been called by every one of the astronomers up to this date *nebulous*, are groups of small stars set thick together in a wonderful way, and although each one of them on account of its smallness, or its immense distance from us, escapes our sight, from the commingling of their rays there arises that brightness which has hitherto been believed to be the denser part of the heavens, able to reflect the rays of the stars or the Sun.

These observations did not immediately prove that the stars were set at varying distances from the Earth, as Digges had depicted them: nothing could do that, except the discovery that different stars display different amounts of parallax, and this was not shown until the nineteenth century. One could, if one chose, think of Galileo as scrutinizing with his telescope the inner surface of the stellar firmament. But as the picture was gradually built up, the conclusion that the fainter stars were more distant became more and more natural: thus the 'sphere of fixed stars' was never directly rejected, but gradually faded out of men's minds.

The last of Galileo's observations was the most exciting. According to Copernicus, the planetary system did not have a *single* centre: the planets went round the Sun, he said, but the Moon went round the Earth. This alleged double-centredness of the planetary system struck some people as the gravest objection to the new theory: if the Moon had not presented this untidy exception to the general rule, they might have accepted the heliocentric view. Galileo now produces from his observations

a notable and splendid argument to remove the scruples of those who can tolerate the revolution of the planets round the Sun in the Copernican system, yet are so disturbed by the motion of one Moon about the Earth, while both accomplish an orbit of a year's length about the Sun, that they consider that this theory about the Universe must be upset as impossible.

The crucial discovery is his observation of the four largest satellites of Jupiter.

> For now we have not one planet only revolving about another, while both traverse a vast orbit about the Sun, but our sense of sight presents to us four satellites circling about Jupiter, like the Moon about the Earth, while the whole system travels over a mighty orbit about the Sun in the space of twelve years.

He reports how on 7th January 1610 he noticed three small bright stars near the planet Jupiter.

> And although I believed them to belong to the number of the fixed stars, yet they made me somewhat wonder, because they seemed to be arranged exactly in a straight line, parallel to the ecliptic, and to be brighter than the rest of the stars, equal to them in magnitude.

He made a note of their positions, and thought no more about them for the moment.

> I scarcely troubled at all about the distance between them and Jupiter, for, as I have already said, at first I believed them to be fixed stars; but when on January 8th, led by some fatality, I turned again to look at the same part of the heavens, I found a very different state of things. . . .

The whole pattern of Jupiter and the stars was now different!

> My surprise began to be excited, how Jupiter could one day be found to the east of all the aforesaid fixed stars when the day before it had been West of two of them; and forthwith I became afraid lest the planet might have moved differently from the calculation of astronomers, and so had passed those stars by its own proper motion.

In the course of the next few nights, he realized that the little stars in question were travelling across the sky always in company with Jupiter:

> Since they are sometimes behind, sometimes before Jupiter, at like distances, and withdraw from this planet towards the east and towards the west only within very narrow limits of divergence, and since they accompany this planet alike when its motion is retrograde and direct, it can be a

matter of doubt to no one that they perform their revolutions about this planet. . . . The satellites which describe the smallest circles round Jupiter are the most rapid, for the satellites nearest to Jupiter are often to be seen in the east, when the day before they have appeared in the west, and contrariwise. Also the satellite moving in the greatest orbit seems to me, after carefully weighing the occasions of its returning to positions previously noticed, to have a periodic time of half a month.

These 'Medicean planets' (as he named them, in honour of the Medici, the ruling house of Florence) made it impossible any longer to argue that all the bodies in the planetary system must be revolving around a single centre. A major objection to the Copernican view was overcome.

In later years, Galileo offered more evidence to support his conclusions: for instance, as a result of his observation of sunspots. Yet nothing which he published afterwards had quite the same universal impact as the *Starry Messenger*. The book became known throughout Europe immediately, and made a special impression in England. Within five years of publication, it was being discussed as far away as Peking. From now on, professional astronomers and mathematicians had their most serious doubts about the Copernican system removed; while, even for amateurs, astronomy became a live subject and telescopes fashionable toys. John Donne might be inclined, on the whole, to remain a supporter of Tycho's compromise-system, and John Milton (in the section of *Paradise Lost* quoted as the motto of the present book), fell back on Osiander's thesis that the business of astronomy was, not to aspire to cosmological truth, but simply

> how build, unbuild, contrive
> To save appearances.

Yet even Milton went out of his way, on a trip to Italy, to beg an interview with the disgraced Galileo, and took the first opportunity to look through a telescope for himself at the Moon and the Heavens. The microscope and the telescope between them provided sources of stimulus and imagery which poets throughout Europe—and particularly in England—took up with enthusiasm and put to good use.

As everyone knows, the story of Galileo's life had an unhappy ending. His famous dialogue on *The Two Chief Systems of the World*, published in 1632, was a comprehensive survey of the points at issue between the Ptolemaic and Copernican views. He presented the full case for Copernicanism in a popular and striking manner, and met the objections against it with arguments, some of which were borrowed from Aristarchos, Oresme, and Copernicus, while others were based on his own work. The book had originally been composed with the encouragement of the ecclesiastical authorities, for Galileo was a good Catholic and had no wish to do the Church an ill-service. For their part, the members of the liberal party within the Vatican would have been happy to see the dispute within astronomy settled without any further scandal. But the dialogue, when it finally appeared, was too uncompromising to be generally acceptable: the conservatives, who included some of Galileo's most influential enemies, succeeded in getting the policy of the Church reversed. Galileo found himself in a position much like that of the aged Pasternak in the Soviet Union, disgraced and under surveillance, and his final *Discourses* on mechanics had to be published in 1639 in Protestant Holland. The story of this whole wretched episode has been admirably told in G. de Santillana's book *The Trial of Galileo*.

To sum up: Galileo did two things for astronomy. First, he made the telescope an indispensable instrument for all future observations: for the first time in history, men became aware that there were other heavenly objects besides those visible with the naked eye. This widening of our horizons, which Galileo began, is still continuing. Secondly, by his consistent advocacy of the Copernican system, regarded as a physical truth, he finally established the right which Copernicus had claimed: that astronomy should be something more than a set of mathematical contrivances, 'merely to provide a correct basis for calculation'. Questions about the structure, fabric, and operation of the Heavens once again fell within its scope.

Beyond his work with the telescope, Galileo did not do very much to solve the outstanding problems of celestial physics. Having so many irons in the scientific fire, he never troubled himself much over the detailed intricacies of planetary geometry, and he was content to argue about the Copernican system as a general cosmological position. Nor did he seriously tackle the question, why the Heavenly bodies move in the orbits they do. He hinted that the clue to this might be that

> the Sun, as the chief minister of Nature and in a certain sense the heart and soul of the universe, infuses by its own rotation not only light but also motion into the other bodies which surround it. . . . If the rotation of the Sun were to stop, the rotations of all the planets would stop too.

But he did not see any need to ask for a cause why the planets should travel in closed orbits, rather than in straight lines. For reasons which were convincing enough before Newton's day, he never entirely abandoned the old ideal of circular motion.

JOHANN KEPLER'S ASTRONOMICAL PHYSICS

As always in science, the solution of one problem brought others in its train. If the coherence of the 'world-system' did not lie in its relations with the starry sphere outside, it was necessary to look for a new sort of coherence *within* the system. John Donne in 1611 was already deploring the collapse of the old cosmology under the impact of Copernicus' and Galileo's criticisms:

> The Sun is lost, and th' Earth and no man's wit
> Can well direct him where to look for it.
> And freely men confess that this world's spent,
> When in the Planets, and the Firmament
> They seek so many new; . . .
> 'Tis all in pieces, all Coherence gone;
> All just Supply, and all Relation.

The lifelong, self-appointed mission of Johann Kepler (1571–1630) was to reveal the new, inner coherence of the Sun-centred planetary system. His central aim was to produce a 'celestial physics', a system of astronomy of a new kind, in which the forces responsible for the phenomena were brought to light.

Kepler's career illustrates the motives and hazards of scientific thought to the point of caricature. He started on his investigations fired by the vision of a theory which would make mathematical sense of the number of the planets and their relative distances from the Sun. There must surely (he thought) be some simple mathematical reason which would explain why there were six, and only six planets, and would show how their respective distances from the Sun were connected with one another. He never wholly lost his faith that such a theory could be found. Yet, despite his commitment to this faith, he was not content simply to *assert*: instead, he was prepared to spend his life checking the adequacy of his ideas against the astronomical records, exploring with meticulous doggedness ways in which they could be brought into line. In the process, and almost incidentally, he brought to light the three Laws of Planetary Motion for which he is still remembered: their full significance became clear only a half a century later.

From some points of view, his actual achievements might appear to have had little to do with his aims, so that his latest biographer has called him a 'sleep-walker'. Certainly his ideas about the planetary distances have dropped out of physics entirely, and his conception of the Sun's action on the planets has been swept away. Yet his general contention has been triumphantly confirmed. The shapes and sizes of the planetary orbits, and the speed with which the planets travel, are in fact—as he passionately believed—related together in a mathematical system; and the key to the system—as he guessed—lies in the influence by which the central Sun controls the movements of the planets.

The particular union which we find in Kepler, of far-reaching original speculation, unlimited patience, and scrupulous attention to the facts, is the characteristic mark of the best modern science. By his early twenties, he had accepted the Copernican doctrine, and was firmly convinced—in addition—that the Pythagoreans had been right. There were mathematical relations to be discovered in the world, and these could be expected to make sense of the planetary system. If he was to draw a truly rational picture of the universe, or 'cosmography', the first problem to solve was this problem of the 'mathematical harmony': this was the problem hinted at by the title of his first book, *Mysterium Cosmographicum*, which he published in 1596.

He looked for a law governing the distances first in simple arithmetic, and then in plane geometry, without success. Lastly, he turned to solid geometry, and was struck by an exciting possibility, which seemed to solve both his problems at once. As we saw earlier, Plato's pupil Theaetetus had proved an important theorem; that there are only five regular solids. In this theorem Kepler now

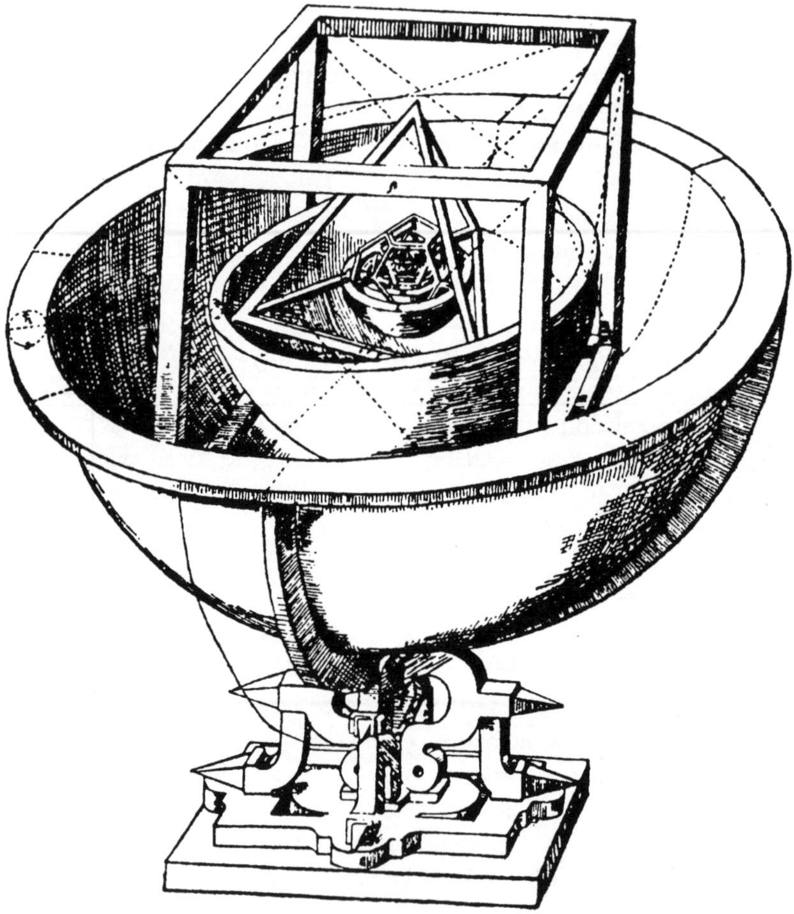

Kepler's model of the universe

The outermost sphere is Saturn's

saw the solution of the 'cosmographic mystery'. If there were only six (known) planets, there were then exactly five interplanetary regions, and he tried to imagine how the proportions of the five Platonic solids might determine the distances between the planets. Taking the greatest and least distances between each planet and the Sun as marking the outer and inner faces of its zone, he built up a mathematical picture of the whole system as a kind of Chinese box. The shell containing the orbit of Mercury fitted inside an octahedron, whose points just touched the inner surface of the shell of Venus. Likewise, between the shell of Venus and the shell of the Earth he fitted an eikosahedron; between the Earth and Mars a dodekahedron; between Mars and Jupiter a tetrahedron; and between Jupiter and Saturn a cube.

This part of Kepler's initial vision has vanished without a trace. We now know of nine planets, and have given up looking for any simple mathematical relationships governing their distances from the Sun. But the theory gave Kepler a framework to direct his research; and during twenty years and more of active work he made three other discoveries of more lasting value. These have more than compensated for any shortcomings in the wider framework.

When Kepler proceeded to study the records, he started asking further questions. He observed that the more distant planets moved slower than the inner ones. Jupiter took twelve times as long as the Earth to get around its track, Saturn 30 times, though the planets certainly did not have 12 or 30 times as far to go. There seemed to him to be only two possible reasons for this. Either each of the planets was kept in motion by its own 'motive spirit', which was feebler the further the planet was from the Sun; and this he found implausible. Or alternatively, some force from the Sun itself must keep the planets moving, its action falling off as the distance increased. The details of Kepler's solar force are not of great importance, for they were later corrected by Newton. What was strikingly original here was the conception itself: the idea that a physical force could originate in the Sun and be the cause of the planetary motions. One cannot find before Kepler any clear recognition that the heavenly motions called for an explanation in terms of a *continuously* acting physical force.

In 1596 Kepler could do little more than put forward these ideas as *possibilities*. To go further, he had to check them in detail against the best planetary records available: those he had access to were not sufficiently reliable. So he got in touch with Tycho Brahe, who had just been appointed as a kind of 'astronomer royal' to the Imperial Court at Prague. Tycho engaged Kepler as a 'research assistant', and the two men embarked on a short and stormy relationship. It lasted from early in 1600 until Tycho's death in October 1601; and during part of the time Kepler was away from Prague.

The problem Tycho gave Kepler was fortunate. Of all the heavenly bodies, the one that gave astronomers most trouble was the planet Mars. Employing, as both Ptolemy and Copernicus had done, an eccentric deferent circle together with a single epicycle, one could in theory keep one's predictions for the other planets in line with observation to an accuracy of one-tenth of a degree, and for the Sun to one-hundredth. Whether Ptolemy or Copernicus ever actually achieved this full accuracy in practice is a matter of doubt, but they had the capacity to do so. Until the time of Tycho, this was also just about the limit of observational accuracy: Copernicus, indeed, was happy if he could get a 'fit' no worse than one-sixth of a degree.

Mars alone was obstreperous. The discrepancy between theory and observation was more than any tinkering could remove. In the case of Mars, the standard 'eccentric-cum-epicycle' construction is bound to lead to errors of a good $\frac{1}{2}°$, and Tycho's painstaking observations put the existence of this discrepancy beyond all doubt. Kepler had now to find an alternative construction, which would bring the observed motion of Mars once more into line with theoretical prediction.

At this point, Kepler's physical genius showed itself. He might have been content to add further circles-on-circles, and to have eliminated the deviations by mathematical artifice. But he was not just an astronomical computer: he had the temperament of a mathematical physicist. The task of 'saving appearances' could have been carried out by pedestrian mathematical means, adding epicycles to epicycles; but this would only have made the cosmographic mystery even more impenetrable. Instead, he was determined to find a simpler solution of the problem—to demonstrate that the orbit of Mars could be represented by *a single geometrical figure*, such as could be explained by the action of simple physical forces.

He thought the task would be completed in eight days: in the end it took him over eight years. He continued to work away at it after Tycho's death, and published the results in 1609 in his *New Astronomy*. This book is a monument to his intellectual passion and dogged persistence. Instead of writing one of those aseptic scientific papers fashionable in the mid-twentieth century, in which the final outcome of the investigation is produced fully formed like a rabbit from a hat, Kepler recorded all his successive attempts to analyse the movements of Mars. One gets from the book, in consequence, much more of the 'feel' of a scientific investigation than a modern scientific journal ever gives.

Before getting around to the *shape* of the orbit, he first asked how the speed of the planet varied at different points in its track. Unlike Copernicus, he did not think that its speed must necessarily be uniform: instead, he experimented with the idea—already present in his earlier book—that the speed of the planet fell off in proportion as its distance from the Sun increased. The idea appeared plausible, and he put forward, as the first principle of his investigation, the mathematical relation which, in its general form, we now know as his Second Law. According to this law, the speed of any planet varies in inverse proportion to its distance from the Sun, in such a way that the line joining the planet to the Sun sweeps out equal areas in equal times (cf. p. 204). Behind this principle lay a dynamical theory which still owed something to Aristotle: he took it for granted that any variation in the Sun's force would produce a proportionate change in *speed*. The Sun, he thought, exerted a sideways force on the planets, so as to keep them moving along their tracks: just as the force exerted by a lever drops off as you go further from the fulcrum, so the Sun's force also must drop off in proportion to distance.

Armed with this principle, Kepler returned to the shape of the orbit. For a long time, he continued to work with the idea that it was, after all, circular; even if this meant giving up his dynamical principles. Eventually, he was forced to try other shapes, since the

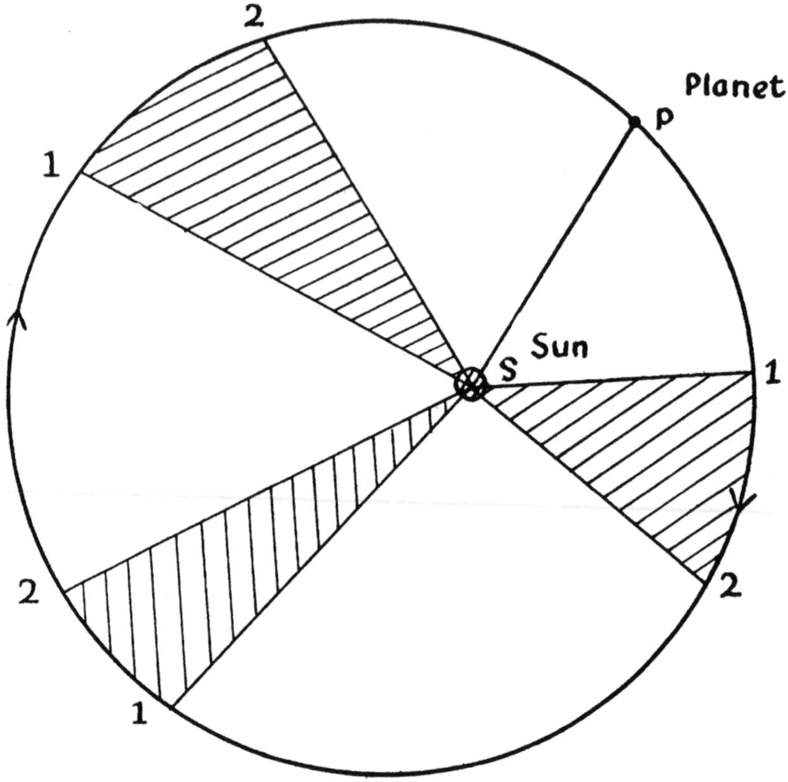

Kepler's second law of planetary dynamics

On Kepler's law, the time taken for a planet to go from point 1 to 2 on the orbit depends on the area of the orbit contained within these points. The line SP will sweep out equal areas in equal times. Notice that when Kepler first formulated this law he did not assume that the orbit was an ellipse; he thought that it was circular, with the Sun placed eccentrically. He realized that, in coming to this law, he had made two questionable assumptions: first, that the sum of an infinite number of lines can be equated with an area, and second, that the orbit was circular orbit seemed undoubtedly to be compressed in one direction. At least once the thought crossed his mind that the shape might be an ellipse, though at first he toyed with this shape simply as a mathematical construction. On physical grounds, he was inclined to believe that the true orbit was egg-shaped. Checking this against the observations, he found some discrepancies still remaining. And, in size, these discrepancies were just equal to the differences between his supposed egg-shaped curve and the ellipse he had used as an approximation. So at last he calculated how Mars would appear from the Earth each night if its orbit was in fact an ellipse, with the Sun at one focus. The last serious discrepancies disappeared.

The story of this untiring investigation has been retold more than once in the last few years. It is easy enough to sum it up in two paragraphs now, when the outcome is so familiar to us. But the sheer labour of eight years' hard calculation, and Kepler's scrupulous honesty in admitting failure over and over again, are impossible to exaggerate.

In the next ten years, Kepler returned to the problem of the number and relative distances of the planets. The theory of the five regular solids was still the starting-point of his investigations, but he was candid enough to admit that the facts did not support his original speculation exactly.

> As regards the proportions between the planetary orbits, the proportion between two neighbouring planetary orbits is always such as can easily be seen to *approach* the proportion of the spheres of one of the five regular solids. . . . Nevertheless, it is not exactly equal, as I once dared to promise.

He now went into the detailed adjustments needed to fit the theory to the facts:

> In short: the cube and the octahedron, which are consorts, do not penetrate their planetary sphere at all; the dodecahedron and eikosahedron, which are consorts, do not wholly reach to theirs, the tetrahedron exactly touches both. In the first case there is a small deficiency, in the second a small excess, and in the third, equality in respect of the planetary intervals.
>
> Evidently then, the true proportions of the distances between the planets and the Sun have not been taken from the regular solids alone. For the

Creator, who is the very source of geometry and (as Plato puts it) 'is forever doing geometry', does not depart from his own specifications. . . . So we must recognise that for working out the diameters and eccentricities jointly, we require more principles, beyond those of the five regular solids.

In his final expositions of Copernican astronomy and 'the harmonies of the world', Kepler went over a great deal of the ground again, but also incidentally presented some new results. He explained his Second Law as a direct application of the principle of the lever:

When the arms [of a lever] are in equilibrium, the ratio of the weights hanging from each arm is inverse to the ratio of [the lengths of] the arms. . . As the short arm is to the long, so the weight on the longer arm is to the weight on the shorter arm. And if we in thought remove one of the arms and, instead of the weight hanging on it, imagine a turning-moment of the same magnitude acting at the fulcrum and keeping the remaining arm with its weight lifted up; then evidently this moment at the fulcrum will not exert the same force on a weight at a distance as it does on the same weight when near. In the same way, astronomy tells us, in the case of a planet, the Sun does not have as much power to move it and cause it to travel round when the planet is further away from the Sun in a straight line as it does when the distance is smaller.

He compared the further cause responsible for each planet varying its distance from the Sun with the force of magnetic attraction:

Is it not unbelievable that the celestial bodies should be like huge magnets? Then read the philosophy of magnetism by the Englishman William Gilbert; for in that book, although the author did not believe that the Earth moved among the stars, nevertheless he demonstrated its magnetic nature by very many arguments. . . . It is by no means absurd or incredible that *all* the chief planets should have the same nature.

At other times, Kepler represented the eccentric motions of the planets by the use of musical notation—a last echo of the Pythagorean 'harmony of the spheres'. He also looked for a mathematical law to govern the relative densities of the different planets:

It is not appropriate that all the planets should have the same material density . . . the chief reason being connected with the differences between their periodic times. . . . Whatever body is nearer to the Sun will also be denser, for the Sun itself is the densest of all bodies in the world, as is evident from its immense and manifold power. . . .

He tried out various possible mathematical relationships, and concluded:

So, reduced to round numbers, the result will be the figures tabulated below; and I find that the terrestrial substances listed beside them correspond fairly closely in ratio—as will be seen in my book written in German in 1616 on the subject of weights and measures:

Saturn	324	the hardest gems
Jupiter	438	loadstone
Mars	810	iron
Earth	1000	silver
Venus	1175	lead
Mercury	1605	quicksilver

thus we may reserve gold, whose density in this ratio is 1800–1900, for the Sun.

(We shall see in our next book how closely this particular idea ties in with the traditional alchemical doctrines. These associated the principal Heavenly bodies with the different substances—the Sun being regularly compared with gold: in alchemical literature, gold regularly had the 'code-name' of Sol, while the planet Mercury still shares its name with the fluid metal, quicksilver.)

Almost buried in these 'harmonic' speculations, we find a brief statement of the third and last of the relations for which Kepler is remembered.

The periodic times of the planets are not in the same proportions as the sizes of their spheres, but greater than them, the exact proportion in the case of the chief planets being the 3/2th power. That is to say, if you take the cube root of the periods of Saturn (30 years) and Jupiter (12 years) and square them, the true ratio of the sizes of the spheres of Saturn and Jupiter will be in proportion. This is so even if you compare spheres which are not neighbouring to one another. E.g. Saturn takes 30 years, the Earth one. The cube root of 30 is about 3·11, that of one is one. The squares of these cube roots are 9·672 and one. So the size of Saturn's sphere is 9·672 times the size of the Earth.

This last relationship

$$(\text{planetary year})^2 \propto (\text{distance from Sun})^3$$

was destined to play a key part in Newton's arguments about gravitation. Kepler had reached it as one by-product, among many others, of his conviction that the planetary system had a fundamentally mathematical structure. The same method also turned up a great deal of intellectual chaff, yet we have no reason to scoff: the fundamental conviction had borne fruit in the form of discoveries which only a man of the highest imagination and persistence could have made.

Kepler and Galileo had reconstructed the cosmic system as completely as anyone could do without an adequate theory of planetary dynamics. In addition Kepler was aware of the need to extend the conception of gravity, which—he saw—was a *reciprocal* attraction:

> The traditional doctrine about gravity is erroneous. . . . Gravity is the mutual bodily tendency between cognate bodies towards unity or contact (of which kind the magnetic force also is), so that the earth draws a stone much more than the stone draws the earth. . . .
>
> If two stones were placed anywhere in space near to each other, and outside the reach of force of a third cognate body, then they would come together, after the manner of magnetic bodies, at an intermediate point, each approaching the other in proportion to the other's mass.

Here was a man who realized that the crucial problem of the planetary system was one of a dynamical balance:

> If the earth and moon were not kept in their respective orbits by a spiritual or some other equivalent force, the earth would ascend towards the moon one fifty-fourth part of the distance, and the moon would descend the remaining fifty-three parts of the interval, and thus they would unite—assuming that they are both of the same density.

But Kepler did not have the means of discovering what the true forces were that kept the Moon and planets in their orbits. The final union of dynamics and astronomy was still sixty years off, and for the moment we must go back to look at the other side of our story.

FURTHER READING AND REFERENCES

The development of astronomy in the period between Copernicus and Newton is discussed in a lively, stimulating but contentious way in the book

Arthur Koestler: *The Sleepwalkers*

For the chief figures in this development, see also

J. A. Gade: *The Life and Times of Tycho Brahe*
Max Caspar: *Johann Kepler* (recently translated into English by C. D. Hellman)
Stillman Drake: *Discoveries and Opinions of Galileo*

For the dispute between Galileo and the Church, see

G. de Santillana: *The Trial of Galileo*
F. Sherwood Taylor: *Galileo and the Freedom of Thought*

The influence of seventeenth-century astronomy on the literature and poetry of the time is a subject in itself. See particularly the pioneer series of essays collected in

Marjorie Nicolson: *Science and Imagination*

For a general account of the development of science during the seventeenth century, see the early part of

A. R. Hall: *The Scientific Revolution*

Sections to illustrate the work of Tycho, Galileo and Kepler are included in the film *Earth and Sky*.

8

The Creation of Mechanics

In astronomy, the half-century from 1570 to 1620 saw a radical break with tradition. In mechanics, there was no such sharp discontinuity. Popular histories of the subject often give Galileo credit for an originality in this field equal to his originality in astronomy. From the time of Aristotle on (they imply) the subject languished, through being left in the hands of monks or philosophers who, with a niggling insistence on drawing distinctions, treated it solely as an exercise in logic-chopping: Galileo, by introducing 'the experimental method', unmasked the elementary blunders passed on from Aristotle by the mediaeval scholastics.

Recent research has shown that this picture is a caricature of the facts. In the centuries between Aristotle and Galileo, mechanics made continuous progress. The Arab mathematicians and philosophers combined the ideas of Archimedes, Euclid, and Philoponos with those of Aristotle, to good effect—particularly in statics. Scholars in Mediaeval Europe took up the problems from an original point of view before the end of the thirteenth century. And Galileo's work can best be understood as the culmination of this mediaeval tradition. The problems he discussed, the methods of argument he employed, the technical terms he used: these were all taken over from the tradition, and his novel discoveries would have been impossible without this background. His experimental discoveries were certainly important, but his most influential contribution may itself have been a literary one: his two full-scale treatises, on astronomy and mechanics, set the problems before the intelligent reader in easily-read Italian dialogue.

Aristotle had given an account of the manner in which bodies move which did rough justice, at the level of common sense, to the most familiar sorts of motion one can see, both on the Earth and in the Heavens. By Newton's time, five important changes had taken place in the theory of motion, and as a result it became possible to produce a new synthesis of dynamics and astronomy. The effects of these changes can be summarized as follows:

(a) Aristotle thought of movement as one special variety—namely, change of position—of the general phenomenon of change. He was interested in movement as a qualitative phenomenon to be explained in the same sort of terms as changes in colour, warmth, or health. In his discussion of movement there was only a minimum of mathematics—simple numerical ratios, for instance, between one distance and another. A mathematical concept such as 'velocity', which is not a simple number but rather a 'dimensional' quantity (a length divided by a time), was something with no place in his system. In Newton, we find the new mathematical apparatus of the differential calculus applied to problems of mechanics.

(b) As Aristotle's discussion of mechanics was qualitative, he drew few of the distinctions which lie at the foundation of modern mechanics. He had no notion of 'instantaneous velocity', and only hinted at a definition of 'acceleration': both ideas were to play central parts in the mechanics of Newton. Also, he worked with a single idea of the bulk of a body, where we distinguish between weight, density, mass, and specific gravity. In the same way, he treated the 'effort' required to move a body sometimes as we would treat 'force', at other times in a way suggesting 'work', 'action', or 'energy'. Finally, his notion of 'resistance' sometimes implied 'density', sometimes 'viscosity', and only sometimes a 'resisting force' in the modern sense. Mediaeval mathematicians gradually gave precision to all these definitions, but the task was not completed until Newton.

(c) Aristotle recognized five different types of material, which moved naturally in quite different ways, according to their location.

After Newton, the distinction between the transitory straight-line motions of terrestrial things, and the unending, circular motion of celestial things, had been entirely broken down. A common set of dynamical principles had been worked out applicable equally to matter of all kinds and in all states.

(d) As this distinction began to break down, it became clear that, under appropriate conditions, bodies on Earth would be capable of continuing in motion for ever, if only they were left entirely to themselves. All that prevented this from happening in practice was the impossibility of removing the hindrances (friction, air-resistance, etc.) which normally bring moving bodies to a stop.

(e) So, at last, men rose to a new and more refined level of mathematical analysis, where the essential relationship was that between force and acceleration. The immediate effect of an outside force was, not to *maintain* the speed and direction of a body's movement, but to *alter* it. Motion conceived as a balance between a motive force and a resistance was given up: in the new picture, movement was something inertial, i.e. naturally self-maintaining. External forces would either slow it down, speed it up, or change its direction.

These five changes were brought together as a result of one last step. The paradigm of uniform circular movement was replaced by that of movement along a Euclidean straight line: the older conception of unhindered, self-explanatory celestial movement gave way to a new mathematical ideal. On this 'explanatory paradigm' of steady rectilinear movement the whole of Newtonian mechanics is founded.

Even as a bare possibility, this new paradigm could hardly have been considered before the year 1600. There were two reasons for this. The very notion of a terrestrial body continuing to move forever is entirely contrary to all everyday experience, and so was foreign to Aristotle's way of thought. It could exist only in a mathematical theory, of a kind which dealt with idealized situations as much as with real-life ones. It is no accident that such a theory was built up only after the renewal of interest during the Renaissance in Plato and Platonic ways of thought.

Secondly: so long as men visualized the whole universe as a closed, finite sphere, a body moving indefinitely in a Euclidean straight line could have no place. Such a body would eventually end up, not only out-of-this-world, but even 'outside Space'. Circular motion was much more appropriate to a spherical universe: introducing the Newtonian paradigm too soon might have delayed, rather than accelerated, the application of the new physics to cosmology. In an infinite universe, such as Digges and Bruno popularized, these objections did not arise.

TREATING MOTION MATHEMATICALLY

Let us now look more closely at each of these points in order; beginning with the ways in which mechanics became quantitative and mathematical. This means studying the development of kinematics during the Middle Ages. Kinematics, as we explained earlier, deals with the movements of bodies in terms of distance, time, and speed alone: it is not concerned with the forces and causes responsible, which form the subject-matter of dynamics. Kinematics is, in fact, the *geometry* of movement, whereas dynamics is the *physics* of movement. Eudoxos and Ptolemy confined themselves to planetary kinematics, Aristotle and Kepler were interested in planetary dynamics also.

The classical Greek discussion of kinematics always took an extremely concrete form. Ships do not have 'a velocity of x m.p.h.': they are 'shifted y feet in z minutes'. Aristotle does, of course, use the words 'faster' and 'slower', but he always specifies speeds in terms of actual distances travelled in given times. Even Archimedes expressed his kinematic theorems in the same terms:

> If some point is moved with a uniform velocity along [the whole length of] a given line, and if we mark out upon this line two [shorter] lines, these will bear the same ratio to one another [in length] as do the periods of time taken by the point in traversing them.

Aristotle twice hinted at the problem of acceleration in the *Physics*, but never satisfactorily defined it. He hardly seemed to think of velocity or acceleration as distinct variables, but only of movements as 'increasing in intensity'. Strato had a first shot at distinguishing an accelerated movement by defining it as one in which equal distances are traversed in successive shorter periods of time: he kept the increments of distance the same. By the 12th century, Jordanus of Nemours was putting the definition the other way around: acceleration occurs when, in equal increments of time, greater and greater distances are traversed. In both these definitions, everything is related back to a constant increment, either a set time or a set distance. The mediaeval mathematicians' first task was to recognize that a body's speed could be treated as a variable in its own right—not just as (say) distance gone in a standard time. Their next problem was to describe the motion of an accelerating body in terms of this property, 'velocity', which might change continuously from one instant to the next.

The changeover to modern mathematical kinematics involved three successive steps, the first two of which were taken before 1400. These were the following: (i) the extension of Aristotle's analysis to cover accelerated motions, (ii) the first use of graphs to demonstrate the overall effect of changes taking place at a variable rate, and (iii) the development of these first simple graphical methods into the 'infinitesimal calculus'.

The first step was almost certainly taken between 1330 and 1350, by scholars at Merton College, Oxford. Several of their manuscripts contain proofs of a theorem which is the key to a satisfactory treatment of uniform acceleration. It looks an extremely obvious statement to us now, for in our most elementary physics we are taught results which in fact depend on this theorem. But it was a definite advance on anything that had gone before. William Heytesbury, in 1335, stated it in terms which can be paraphrased as follows:

> Starting either from zero, or from some finite value, any change in velocity which terminates at a finite value and is acquired or lost at a uniform rate will be equivalent in effect to its mean value. So a moving body, acquiring or losing velocity uniformly during some given period of time will cover exactly the same distance as it would have covered in the same period of time if it had moved steadily at its mean value of velocity . . . namely, at the value which it would have at the middle moment of time.

The total distance travelled by a body in a uniformly accelerated motion will be exactly the same as if it had been moving for the same period at the *mid-time* speed. Does this sound obvious? The result is very far from obvious, and its implications are easily overlooked. For instance: if a racing-car accelerates *uniformly* from rest to 140 m.p.h. over a distance of two miles, what will its speed be half-way, i.e. after the first mile? Off-hand, one would be tempted to say: '70 m.p.h., of course'. This would be to make the very mistake the Merton mathematicians avoided. For, in a uniform acceleration, the car would reach half-speed at half-time, but in doing so would cover only *one-quarter* the distance: by the time it reached the half-way mark in distance, it would be doing close to 100 m.p.h., and the second mile would be covered in a much shorter time than the first.

The Merton school, then, defined acceleration as occurring when, in each successive period of time, a body covered the same *additional* distance more or less. The crux of the matter was the *time*: given this definition, their 'mean speed theorem' could be proved without elaborate mathematics.

They immediately went on to draw a further conclusion; i.e. that a body accelerating uniformly from rest will cover three times as much ground during the second half of the time as it has during the first. And the ways in which they demonstrated this show the emergence of graphical technique as a mathematical method. Nicholas Oresme, around 1360, in his treatise *On the Configurations of Qualties*, showed how to use a primitive graph to anticipate the calculus (cf. p. 216). Suppose we consider a change of any sort, for instance, a movement proceeding at a uniformly increasing rate. Draw a horizontal line ABCD to represent three successive minutes. Next draw a sloping line AEFG to represent the steadily increasing velocity, starting from rest. Now, declared Oresme, the sizes of the triangles AEB, AFC, AGD correspond to the complete changes produced in one, two, three minutes: i.e. in the case of motion, to the total distances gone. Uniformly accelerated motion from rest lasting two minutes gives a triangle whose area is four times that for the same motion over one minute: in the second minute the body will have gone three times as far as in the

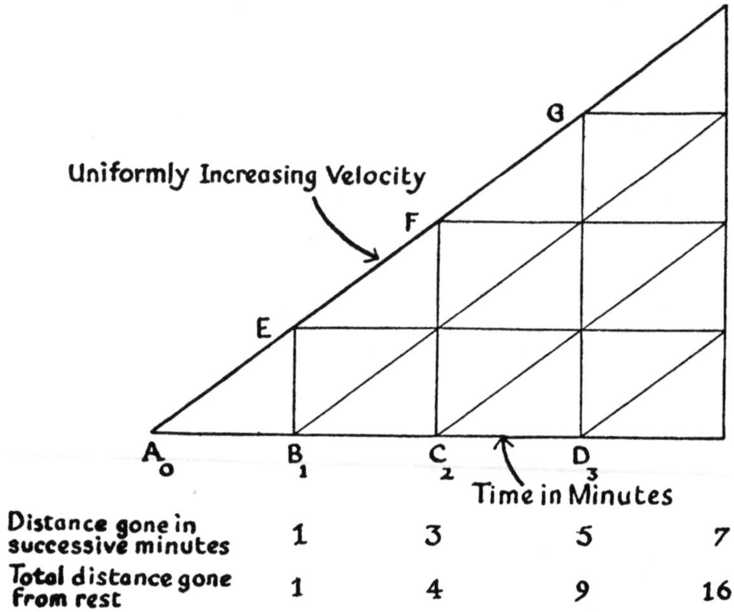

Oresme's method for representing changes graphically

The triangles AEB, AFC, AGD correspond to the total changes in one, two and three minutes respectively. In the second minute (BEFC) the body will have gone three times as far as in the first minute; in the third minute (CFGD) it will have gone five times as far as in the first minute. The changes taking place increase in proportion to the odd numbers

first. And in general, the changes taking place in successive minutes will be in proportion to the successive odd numbers: in our example, the distances gone in the first, second, third . . . minute will be proportional to 1, 3, 5 . . . respectively. From this, the step to the general formula for uniform acceleration

$$(\text{distance}) \propto (\text{time})^2$$

awaited only the introduction of algebraic notation. For the areas of the triangles in Oresme's diagram, which represent the distance, are proportional to the squares of their base-lines, which represent the time.

This kind of graphical procedure provided the impulse for two important steps forward. Firstly, it encouraged men to look for ways of replacing *qualities*, which Aristotle had regarded as fundamental, by 'numerical degrees' or *quantities*. Even the idea of a scale of temperature was foreshadowed—more than 200 years before anyone built a satisfactory thermometer—in the following passage:

> Given a quantity of water of two weights, with a heat of degree six; . . . and another quantity of water of one weight whose heat is of degree 12; suppose the two quantities of water mixed, the hotness of the mixture will be raised to degree 8 on the scale of intensity. . . . since the distance from 6 to 8 is one-half the distance of 8 to 12, just as the water of one weight is one-half the water of two weights.

The other important line of development initiated by Oresme's work on graphs led on to the idea of 'integration' in the modern calculus. For the use of the triangle to illustrate the Mean-Speed Theorem quickly caught on: it reappears, for instance, in Galileo's own treatise. As time went on, diagrams of this sort became more and more sophisticated, eventually taking the form familiar to anybody who has worked through a text-book of the calculus.

By the time we reach the journal of Isaac Beeckman (1618) we are on the very verge of the modern method. Beeckman treats a stone falling in a vacuum as an example of uniform acceleration. He divides the time of fall up into 'indivisible moments'—i.e. infinitesimals—and asks how the distances travelled in these successive moments will contribute to the total motion. Representing the 'indivisible' moments to begin with as of finite size, and thereafter reasoning about them as negligibly small, he gives a mathematical proof reminiscent of Archimedes: the area of the triangle, he shows, will approximate to the total distance gone the more exactly, the smaller we imagine the 'indivisibles' to be. This proof was put on a formal basis by Newton and Leibniz at the end of the century.

Hand in hand with this discussion of acceleration went a progressive refinement in the definitions of the chief kinematic concepts. The mediaeval scholars' practice of drawing precise distinctions was in this case of the greatest value: if it was 'logic-chopping', it was done to good purpose. They clearly distinguished motion, acceleration, and speed, in all their varieties. Oresme, for instance, contrasted linear and circular velocity, and distinguished their respective measures or 'dimensions':

> In circular motion a body is both said 'to be moved' and 'to revolve'. Now the intensity of a linear velocity is measured by the linear distance which will be covered at that degree of speed. But the intensity of a rotational velocity is noted by the angles described about the centre of motion. Hence one body moving in a circle may *travel* more rapidly than another and yet *revolve* less rapidly. Now, for instance, perhaps Mars *travels* more rapidly in its own motion than the Sun, on account of the size of the circle it describes, yet the Sun completes its circle more swiftly and *revolves* more swiftly around the centre. . . . Astronomers in practice pay more attention to the rotational [angular] velocity than to the [linear] velocity of motion.

Still, although many of these vital distinctions were made early in the fourteenth century, nevertheless one finds confusion about some of them recurring right up to the time of Galileo. Galileo himself, in 1604, was still capable of falling into our racing-car fallacy. The ideas finally crystallized in men's minds only after two further things had been done: when, first, all these scattered theorems had been brought together into a single mathematical exposition and presented as a coherent system of propositions; and when, secondly, real-life systems were identified which could be taken as concrete examples—and reminders—of the different species of motion which the mediaeval mathematicians had discussed in the abstract.

Galileo's work on mechanics effectively displaced that of his predecessors, just because he did these two things. After he had published his dialogues, it was no longer necessary to refer back to the centuries of original work which lay behind them. He did for mediaeval mechanics what Euclid had done for Greek geometry and, from 1640 on, the work of the Merton and Parisian Schools could be—and was—quietly forgotten. Furthermore, he produced at least one memorable demonstration of a body actually performing 'uniform acceleration': namely, a ball rolling down an inclined plane.

This demonstration is described in a famous passage in his *Discourses on Two New Sciences*. Galileo's spokesman in the dialogue has explained why one should define uniform acceleration, in the way the Mertonians had done, as 'motion in which velocity increases by equal amounts in equal times'. One of his listeners then asks for a practical illustration of the point. The example given is the following:

A piece of wooden moulding or scantling, about 12 cubits long, half a cubit wide, and three fingerbreadths thick, was taken; on its edge was cut out a channel a little more than one finger in breadth; having made this groove very straight, smooth, and polished, and having lined it with parchment, also as smooth and polished as possible, we rolled along it a hard, smooth, and very round bronze ball. Having placed this board in a sloping position, by lifting one end some one or two cubits above the other, we rolled the ball, as I was just saying, along the channel, noting, in a manner presently to be described, the time required to make the descent. We repeated this experiment more than once in order to measure the time with an accuracy such that the deviation between two observations never exceeded one-tenth of a pulse beat. Having performed this operation and having assured ourselves of its reliability, we now rolled the ball only one-quarter the length of the channel; and having measured the time of its descent, we found it precisely one-half of the former. Next we tried other distances, comparing the time for the whole length with that for the half, or with that for two-thirds, or three-fourths, or indeed for any fraction; in such experiments, repeated a full hundred times, we always found that the spaces traversed were to each other as the squares of the times, and this was true for all inclinations of the plane, i.e. of the channel, along which we rolled the ball. We also observed that the times of descent, for various inclinations of the plane, bore to one another precisely that ratio which, as we shall see later, the Author had predicted and demonstrated for them.

For the measurement of time, we employed a large vessel of water placed in an elevated position; to the bottom of this vessel was soldered a pipe of small diameter giving a thin jet of water, which we collected in a small glass during the time of each descent, whether for the whole length of the channel or for a part of its length; the water thus collected was weighed, after each descent, on a very accurate balance; the differences and ratios of these weights gave us the differences and ratios of the times, and this was with such accuracy that although the operation was repeated many, many times, there was no appreciable discrepancy in the results.

It is important to see clearly what this demonstration proves, and what it does not prove. Galileo *was not* proving that 'a body accelerating uniformly from rest travels a distance proportional to the square of the time'. That conclusion follows directly from the definition of 'uniform acceleration': it is a *logical consequence* of the definition, and *experimental evidence is not required to reinforce it.* On the other hand, Galileo *was* showing that motion of a familiar kind exemplifies with great accuracy the 'uniform acceleration' that previous scholars had defined and discussed. He was further convinced that what had proved true for a ball on an inclined plane would be equally true of a ball falling through a vacuum: the effect of the incline was only to slow down the motion. Notice from his report, how he varied the slope of the plane to check that the acceleration remained uniform for steeper, as well as for less steep, inclines. If the incline had been vertical, he concluded, the ball would still have fallen in the same general way—accelerating uniformly.

This experimental demonstration of Galileo's perfectly illustrates the particular—and original—direction which he gave to mechanics. Whereas in astronomy his discoveries helped to break down the traditional cosmology, in mechanics (as in some other branches of physics) his role was, not to destroy the mediaeval tradition, but to fulfil it. Mediaeval scholars such as Heytesbury and Oresme had proved that, *if* a body were ever to accelerate from rest uniformly, *then* its distance from the starting-point must by definition increase in proportion to the square of the time: Galileo demonstrated that, to as high a degree of accuracy as he could measure, a ball rolling down a smooth inclined plane behaved in just that way. Similarly in the study of heat: where fourteenth-century 'natural philosophers' saw what conditions must be satisfied by an adequate scale of 'warmth-measurement', Galileo, the seventeenth-century 'physical scientist', actually built a practical thermometer in order to get numerical estimates of warmth.

Galileo's celebrated invention of the 'experimental method' was, in this way, the instrument not so much for refuting mediaeval errors, as for showing the relevance, to the real world, of ideas which hitherto had been developed only in the abstract. In its theoretical content, the mediaeval tradition in mechanics, up to and through Galileo, is continuous.

MOTION AND FORCE

So far we have, ready for Newton, three results: a useful and consistent definition of acceleration, the beginnings of the calculus, and a new insight into the geometry of free fall. But this is all within the limits of *kinematics*, and tells us nothing about the forces and causes of motion. We must now see how, in the centuries before Galileo and Newton, the outstanding *dynamical* problems were tackled. We need to look briefly at two problems: the cause of projectile motion, and the influence of forces and resistances on a body's speed.

Aristotle had never explained very convincingly how a body could continue to move, when it was no longer being forcibly pushed. This was the famous problem of 'projectiles'. Philoponos had introduced the idea of 'incorporeal moving powers' which, having been imparted to a body, were responsible for its continuing in motion; and this idea was considerably developed by the Arabs. The first mediaeval writer to adopt this view was John Buridan, a Parisian scholar and Oresme's teacher, working around 1330. Buridan thought of the 'incorporeal power' as though it were an inner energy, whose continued activity was the cause of the body's continued motion: he called it 'impetus'.

There were two contrasting opinions among mediaeval scholars, as there had been among the Arabs. Either impetus was a kind of unnatural *addition* to the normal properties of the moving body, which would tend, as time went on, to die away and disappear of its own accord; or it was capable of lasting indefinitely, so long as nothing interfered with it. The first view compared a moving projectile with a bell set vibrating by a blow. In the same way that the sounds emitted by a bell will progressively die away, so the impetus of the projectile could be expected to disappear. Impetus was self-expending. (This was Oresme's opinion.) Buridan himself thought otherwise: impetus, in his opinion, was 'remitted' or 'corrupted' only by outside interference or resistance—as though a bell would go on ringing for ever, unless its vibrations were forcibly muffled.

Buridan rejects the ideas quoted by Aristotle as possible explanations of projectile motion. If air rushed round to the back of a projectile to fill up the vacuum and so kept it in motion, then a lance 'streamlined' at both ends would stop more quickly than one offering at the back a greater surface to the air. In any case, there was more direct evidence against the theory:

> A ship being pulled along a river, even against the current, cannot be quickly stopped after the pulling has ceased, but continues to move for a long time. Yet a sailor on deck does not feel any air pushing him from behind. He feels only the air from the front resisting. And suppose that the ship were loaded with grain or wood and that a man were situated behind the cargo. Then, if the air were of such force that it could forcibly push the ship along, the man would be squeezed violently between the cargo and the following air. Experience shows this to be false. . . .
>
> Thus we must say that, in a stone or other projectile, something is impressed which acts as the motive force of the projectile . . . in the direction towards which the original mover was moving the body: up or down, sideways or in a circle.

Buridan estimates impetus as being proportional to the speed of a body and its quantity of matter.

Notice that Buridan thinks of impetus as a continually active *cause*: not as an effect or a measure of the motion.

> Impetus is not the very movement itself by which the projectile moves, because impetus causes the motion of the projectile, and a change must be produced by an agent. So it is the impetus which is responsible for the motion, and the same thing cannot produce itself.

Thus, a freely falling body is being acted on by two causes, gravity *and* impetus:

> From this theory we can also see why the natural motion of a heavy body downwards is continually accelerated. At the beginning, gravity alone caused the motion and therefore moved the body more slowly; but as it did so it impressed an impetus on the body. So, now, the motion is caused by this impetus together with the gravity. In this way, the motion becomes faster; and in proportion as the speed increases, so the impetus becomes more intense.

Buridan even applies his theory of the impetus to celestial dynamics. The 'intelligences', or angelic powers, which had been allotted the task of keeping the Heavenly spheres in motion had, as Buridan pointed out, no authority in the Scriptures; and the impetus theory makes them unnecessary:

> One might reply that God, in creating the world, set each of the celestial orbs in motion as He chose, and in doing so impressed on them impetuses which kept them in motion without his having to act on them any more... And these impetuses which He impressed on the celestial bodies were not decreased or corrupted afterwards, because those bodies had no tendency to move in any other way, nor was there any resistance capable of corrupting or repressing the impetus.

Buridan did not put forward this view at all dogmatically, since there might have been theological objections that he had not recognized. Still, he had pointed a way towards overcoming any *absolute* distinction between the motion of Heavenly and Earthly things. One might see, in the continued motion of projectiles, a hint of how Heavenly bodies continue to circle round the Earth without any loss of speed. Once God had set the planetary orbs spinning at the Creation, they would spin indefinitely, like a frictionless potter's wheel.

There matters rested, until shortly before Galileo. In 1585 Benedetti used the impetus doctrine to explain the motion of a stone released from a sling: where Buridan's impetus might be circular, Benedetti's carried the stone off along a straight line.

> A heavy body is projected further by a sling than by hand because, as the sling revolves, a greater impetus of motion is impressed on the heavy body than happens with the hand alone. When the body is freed from the sling, and moves under its own nature alone, it follows a trajectory from the point of release tangential to the last revolution of the sling ... under the influence of the impetus already set up in it.

Yet the traditional ideal of circular motion kept its attractions, and Galileo himself never entirely abandoned it. Buridan had argued that the planets might continue moving for ever, just because they were completely unresisted. Galileo now demonstrated that, under sufficiently similar circumstances, terrestrial bodies would behave in the same way: i.e. they would move unflaggingly *and* in circles. He took another inclined plane and this time rolled a ball *up* it. When the slope was steep, the ball travelled only a short distance before stopping; when the incline was more gradual, the ball went much further before coming to rest. The distance it ran up the slope depended solely on the speed it started with and the angle of the slope. For a given speed, there was a definite vertical height through which the ball would rise.

But what if the plane were horizontal? In that case, the ball could never rise to the height at which it would normally stop: Galileo concluded that friction and air resistance alone prevented it from rolling on for ever without loss of speed. In his youth he had accepted Oresme's doctrine that impetus was self-expending, but now he became convinced that it could be permanent. Furthermore, he now treated impetus, not as the *cause* of continuous motion, but simply as a *measure* of the motion.

Still, a 'horizontal plane', as Galileo very properly recognized, is a surface having at every point the same height above sea-level: it is in fact a sphere. When Galileo describes *large-scale* inertial movement, it is obvious that this is his conclusion:

> Supposing all external impediments were removed, a heavy body on a spherical surface concentric with the Earth would be neutral between rest and movement towards any part of the horizon whatever, and would maintain whatever state of motion it was originally started with. Thus, if put at rest originally it would remain at rest, and if put in motion towards the West (say) it would maintain the same motion. A ship, for instance, once having acquired some impetus in a calm sea, would continue to move around our Earth without ever stopping, or if placed at rest would remain perpetually at rest—supposing in the first place that one could remove all external impediments, and in the second that no external cause of motion supervened.

So, for Galileo, inertial motion is still circular motion. He does not talk of his ship as being prevented by gravity from moving off along a line at a tangent to the Earth. The first essential step towards Newtonian inertia had been taken: terrestrial bodies could in theory keep moving for ever, just as much as celestial ones. But the second step, to the Euclidean straight-line ideal, was left to his successors.

The one real discontinuity in the development of mechanics arose over the question of force and resistance. Galileo had entirely re-analysed the geometry of motion, putting dynamical questions aside. The result of his kinematical work was to show that *none* of the traditional accounts of the relation between force, speed and resistance was satisfactory. The whole problem had to be restated, and in the process the question was stood on its head. The old question had been: 'What force keeps a body moving?' In the light of the newly-recognized tendency of bodies to go on moving indefinitely, it was now necessary to ask instead: 'What force causes a body to stop or otherwise alter its motion?'

To talk of a discontinuity here is not to imply that dynamics had stood still since Aristotle. There were in fact, during the Middle Ages, three different views about the way in which the motive force and resistances acting on a body combine to determine its speed. Aristotle's theory laid down that the speed varied in proportion to the ratio of force to resistance: his chief difficulty had arisen in the case of a vacuum. In the absence of any resistance, a body would on his view travel from one place to another in *no* time. This problem was met by Philoponos with the suggestion that the crucial thing was the *difference* between force and resistance (force minus resistance) rather than their *ratio* (force divided by resistance): on this view the speed of a body in a vacuum would be directly proportional to the force. Thomas Bradwardine, in the fourteenth century, offered a compromise between these two older views more complicated than either.

All three views had attractions, but the whole issue was clouded by the ambiguities in the terms 'resistance' and 'force'; Galileo himself, after inclining to Philoponos' opinion during his days at Pisa, came to see how complicated the whole question was, and steered clear of it.

THE NEW IDEAL: STRAIGHT-LINE MOTION

In Aristotle's tidy spherical universe, circular motion had appeared entirely 'natural'. But the new vision of an infinite cosmos opened up by Digges and Bruno cleared the way for a quite novel conception:

that straight-line, not circular, motion was the proper ideal for physics. This new view was made explicit by René Descartes (1596–1650). He declared that God had created a definite amount of rectilinear motion in each direction, so that a body could not change its direction of movement except by acquiring motion from, or losing it to, another body.

Once men started to think of motion as taking place within an unlimited Euclidean framework, the new view began to seem, in its turn, 'only reasonable'. After all, even circular motion was a deviation from a straight line in one direction or another; and, surely, it was necessary to explain why a body moved away along its circular track on this side rather than that. So what of Galileo's ship? Unless the Earth was to retain its old, privileged place as a 'natural' centre of circular motion, some physical force must be keeping the ship on the surface of the sea. This force was, fairly clearly, its own weight; and Galileo's pupils accepted this view as a natural extension of their master's doctrines.

Descartes put forward the idea of straight-line inertia on metaphysical grounds, as part of a completely geometrical picture of the world; not as the result of any mechanical experiments. But it was soon applied by others to more practical problems. Christiaan Huygens (1629–1695) used it to work out the first satisfactory theory of centrifugal force. Benedetti had discussed the example of a sling, which prevents a stone from flying off at a tangent, and constrains it to move in a circle. Now, argued Huygens, the natural tendency of the stone is to move off in a straight line, and this tendency is more and more forcible, the tighter and faster the circles in which the sling is moved. The explanation lies in the fact that the stone is being continually accelerated. It is deflected at every moment round and round the circle by an amount proportional directly to the square of its velocity, and inversely to the radius of the circle. Since this acceleration was being produced by the sling, the force it exerted to prevent the stone flying away presumably increased in the same ratios as the acceleration. Here we see the first systematic application to physics of the new conception: that the force required to move a body, resistances apart, is proportional to its acceleration rather than its velocity.

Soon, men began to apply this idea not only to terrestrial motions such as that of a stone in a sling, but also to the Heavens. Borelli (1608–1679), for instance, applied Huygens' formula to the planets: he concluded, as a result, that the force exerted by the Sun on the planets must be an attraction. Huygens further demonstrated that the attraction would be an inverse-square one, in order to account for Kepler's Third Law—*if* the planetary orbits were all circles. The question outstanding was: did the same hold true of elliptical orbits? This was a real step forward from Kepler's own theory, that a continuous lateral force from the Sun was required to keep them moving along their orbits. The scene was now set, and the characters in place, for Newton. (Plate 13.)

FURTHER READING AND REFERENCES

The development of dynamics and kinematics in the mediaeval period has been studied afresh during the last few years, and the results of this scholarly work have now been brought together in a wholly admirable and impartial summary

M. Clagett: *The Science of Mechanics in the Middle Ages*

Particular mediaeval writers, such as Bradwardine and Heytesbury, have been the subject of monographs published in the series of Studies in Mediaeval Science issued by the University of Wisconsin. For Galileo and his immediate predecessors, see

A. Koyré: *Etudes Galiléennes*
Lane Cooper: *Aristotle, Galileo and the Tower of Pisa*

For the transition to the Euclidean view of space, see

A. Koyré: *From the Closed World to the Infinite Universe*
C. C. Gillispie: *The Edge of Objectivity*

9

The New Picture Takes Shape

THE central theme of Tolstoy's *War and Peace* is the role of the individual in history. Only in our immediate private lives, Tolstoy argues, have we any real power to influence and initiate events. Soldiers and statesmen think of themselves as controlling the great forces of history by which the outcome of political or military events are determined; but in this belief they are perpetually deluded. Rather, the forces of history control them: all they can do is to play out a predetermined part on the stage which is already prepared for them. The acts of the historical drama will follow one another in the same general sequence regardless what actor takes any particular part. If the principal drops out, there will always be an understudy.

Tolstoy's question arises in the history of science also. Do the men of any generation take just those steps which, in the intellectual climate of their time, are 'ready to be taken'? Or can they leap far ahead of their times, forcing thought along genuinely new and creative paths? These questions provide the occasion for plenty of fascinating speculation—especially in the case of a man like Isaac Newton (1642–1727). Looking at his work from a distance, one may get the impression of a single individual precipitating an intellectual landslide: creating single-handed a complete new synthesis of dynamics and planetary theory. Then, coming closer and looking at the detailed problems he tackled in the context of the time, one may go to the other extreme: so many of the pieces united in his jigsaw-picture were about to fall into place of their own accord, that Newton seems like a man who did those things, and precisely those things, which history required him to do.

Actually, the history of science throws a particularly revealing light on Tolstoy's problem. Whether he is a politician or a scientist, a man can do in his own time only a job which is there to be done; but there may be many different *ways* in which he can do it. Even if Newton had never lived, the individual steps that he took would, without much doubt, have been taken by others before the year 1750. What distinguished Newton's work was, not its details, but its comprehensiveness. Even in England there were a good half-dozen men in his time who shared his background, and had access to the same material. Yet none of them saw more than a part of the pattern. Only Newton had the imagination and the mathematical capacity, first to conceive the whole system of ideas, and then to demonstrate that it alone could pull the disunited strands together, and weave them into a single fabric.

THE MAN AND HIS TASK

Different natural sciences demand of their great men different personal qualities: in some there is more scope for speculative imagination, in others for mature reflectiveness. In certain fields, a man may achieve his deepest insights only in later life, at the age of fifty or more. In mathematics and theoretical physics, his best work is often done before thirty. Like all simple generalizations, this one is subject to qualifications, yet it has a basis in experience, which is reflected in the ages at which various classes of scientist achieve election to the Royal Society. So it is interesting to notice that, according to his own account, Newton laid the foundation of his great system of celestial physics at the age of twenty-three.

He had been born in 1642, the year of Galileo's death. (The astronomical prodigy Jeremiah Horrocks, who was a keen disciple of Kepler, had died at twenty-four in the previous year.) Newton grew up at Woolsthorpe, near Grantham in Lincolnshire, and went to Cambridge as a student in 1661. There he studied under the mathematician, Isaac Barrow, whom he was to succeed as Professor of Mathematics. Because of the Great Plague of 1665, the students were dispersed, and Newton spent this enforced vacation at Woolsthorpe. It was during this period of solitary thought that the essential ideas came to him, which were eventually to form the heart of his system. Shortly after the university reassembled, he was elected a Fellow at Trinity College, and two years later succeeded to Barrow's Chair. He kept this position until 1701, also holding in his later years the office of Master of the Mint. From 1703 onwards he was President of the Royal Society.

A young man of Newton's brilliance would, in these days, be taught to take for granted an obligation to develop and apply his scientific talents to the utmost, and to publish his results. Newton recognized no such obligation: for him, 'natural philosophy' was an intellectual pastime—like the highest grade of crossword-puzzle. Once he was satisfied that he knew the solution of a physical problem in outline, he turned to what were for him more serious matters and, as he wrote to Robert Hooke in 1679, 'grutched the time spent in that study unless it be perhaps at idle hours sometimes for a diversion'. Only in the mid 1680's, in response to the enthusiastic interest of his friend Edmund Halley, did he set about preparing a systematic account of his discoveries. (The Royal Society's funds being at low ebb, Halley also bore the heavy expense of publishing, and the book finally appeared in 1687, with the title, *Philosophiae Naturalis Principia Mathematica.*)

Some historians have seen a further reason for Newton's slowness to publish in his stormy relations with Hooke, who supposedly irked him by his jealousy and criticism. Whether this is true or no, it is at any rate clear that for Newton the satisfaction of physical discovery was purely personal and intellectual. He did not share the passionate faith of Francis Bacon in the promises of scientific technology for the betterment of human life.

By this stage in our story, we are already familiar with most of the raw material which Newton used for his new picture. We must now see how he sifted it, and what he made of it. The first essential ingredient was Galileo's work on terrestrial inertia. All Newton's work began from the idea that, for a body left to itself, a state of uniform rectilinear motion is neither more nor less 'natural' than a state of rest. This, as we have seen, was a step beyond Galileo's own conclusions. For Galileo had still thought of 'inertial motion' as naturally circular, whether in the case of a ship moving over the

Earth's surface, or in that of a satellite circling around the planet Jupiter.

Newton himself followed Descartes in accepting the ideal or Euclidean straight-line motion. The idea of Euclidean space was fundamental. He could not allow that any particular object—whether the Earth or the Sun—had a privileged position in space, or a defined 'natural place', in the way the centre of the Earth had done for Aristotle. So far as the laws of motion were concerned, all frames of reference moving uniformly with respect to one another were equivalent.

Having been too quick to accept circular motion as natural, and having excluded discussions of force from his theory of mechanics, Galileo left two separate groups of problems to be solved. On the one hand, a force would be needed to keep his idealized ship circling around the Earth from flying off at a tangent—as, on Descartes' view of 'natural' motion, it should do. Again, forces would also be needed to explain why the planets and satellites moved along closed orbits. Descartes explained the circular motion of celestial things in terms of 'vortices': on his theory, the whole of space was filled by a thin fluid of corpuscles, containing circulatory eddies, whose motion carried the planets and satellites around massive bodies. These vortices were maintained, as eddies in rivers are, by the continual transfer of momentum between colliding corpuscles. Newton rejected Descartes' vortex theory of the planetary forces, but agreed that deviations from Euclidean straight-line motion implied the action of forces of *some* kind. If bodies moved towards the centre of the Earth or the Sun rather than to any other points in space there must be some *physical* reason for this: whenever a body curved away from a straight-line path, some force *must* explain why it curved away in the direction it did.

The forces required to explain these two circular motions, in the Heavens and on Earth, might be quite separate and independent. But then again, they might be related: if so, the problem was to extend Galileo's analysis of terrestrial motion to the motion of Heavenly bodies also. Newton found his clue in Kepler's discoveries. Disregarding Kepler's attachment to the Platonic solids, his views on solar magnetism and the motive spirits, Newton made use of three ideas—the 'three laws' for which Kepler is still remembered.

It was no accident that, for Aristotle, the boundary between the Earth and the Heavens had been the sphere of the Moon—everything one could observe beyond the Moon displaying a free circular motion, everything on this side gravitating with respect to the centre of the Earth. For the Moon is in fact the most distant visible object whose motion is predominantly controlled by the gravitational field of the Earth. The Moon was, therefore, the crucial case for Newton also. If the Moon's orbit could be explained in terms of ordinary terrestrial forces and principles, this could be the stepping-stone. One might then succeed in generalizing the explanation, and apply it to the satellites moving round Jupiter, and to the six regular planets moving round the Sun.

Everyone knows the story told by Newton's friend, William Stukeley, that the key-idea of gravitation came to Newton when he saw an apple fall in the orchard at Woolsthorpe. This is a story which, unlike many legends, makes scientific sense—with suitable qualifications. One must not think of the apple as falling vertically; as though nobody knew that heavy bodies 'gravitate'. One must think, rather, of the apple-trees as shaken by a wind, which tears the fruit loose from the branches and sends them falling sideways, to land, some further, some less far from the tree. The imaginative stroke was to ask: 'What would happen if the apples started their fall even faster? If they were going fast enough, their natural tendency to fly off at a tangent would be balanced exactly by their tendency to fall towards the ground. Air resistance apart, they might travel round the Earth in a complete loop. And, supposing the Moon was travelling free of all resistances, one might perhaps account for its motion in a similar way.'

NEWTON'S ARGUMENT

To see how Newton built his new picture of the Heavens and the Earth from these different ideas and clues, let us follow through the main points of his argument, as he set them out himself in a popular exposition called *The System of the World*.

He starts by making two assumptions which his subsequent argument will in due course confirm, but which for the moment are hypothetical. These are (i) the Copernican doctrine, that the Earth is a satellite of the Sun, like the other planets, and (ii) the thesis that inter-planetary space is effectively empty. Rather fancifully, he presents these two views as hallowed by antiquity in a scientific Garden of Eden, before men were led astray by the serpents of Greek philosophy.

> It was the ancient opinion of not a few, in the earliest ages of philosophy, that the fixed stars stood immovable in the highest parts of the world; that under the fixed stars the planets were carried about the Sun; that the Earth, as one of the planets, described an annual course about the Sun, while by a diurnal motion it was in the meantime revolved about its own axis; and that the Sun, as the common fire which served to warm the whole, was fixed in the centre of the Universe. . . . It was agreed [also] that the motions of the celestial bodies were performed in spaces altogether free and void of resistance. The whim of solid orbs was of a later date, introduced by Eudoxus, Calippus, and Aristotle; when the ancient philosophy began to decline, and to give place to the new prevailing fictions of the Greeks.

In any case, he argues, there is positive evidence against the solidity of the planetary spheres:

> The phenomena of comets are quite inconsistent with the idea of solid orbs. The Chaldeans, the most learned astronomers of their time, looked upon the comets (which of ancient times before had been numbered among the celestial bodies) as a particular sort of planet. . . . And as it was the unavoidable consequence of the hypothesis of solid orbs, while it prevailed, that the comets should be thrust down below the Moon; so, when later observations of astronomers [e.g. Tycho] restored the comets to their ancient places in the higher heavens, these celestial spaces were at once cleared of the incumbrance of solid orbs.

He then raises the question, why the planets are 'retained within certain bounds in these free spaces'—i.e. travel in closed orbits—and are 'drawn off from the rectilinear courses which, left to themselves, they should have pursued'. Kepler and Descartes, he said, 'pretend to account for it by the action of certain vortices'; while 'Borelli, Hooke and others of our Nation' do so 'by some other principle of impulse or attraction'. He himself, in order not to be drawn into a fruitless discussion about the nature of gravitational force, intends to demonstrate only 'in a mathematical way' what form of law must be followed by 'the centripetal force' responsible for the planetary orbits.

Straightway he presents his key thought:

That by means of centripetal forces the planets may be retained in certain orbits, we may easily understand, if we consider the motions of projectiles; for a stone that is projected is by the pressure of its own weight forced out of the rectilinear path, which by the initial projection alone it should have pursued, and made to describe a curved line in the air; and through that crooked way is at last brought down to the ground; and the greater the velocity is with which it is projected, the farther it goes before it falls to the Earth. We may therefore suppose the velocity to be so increased, that it would describe an arc of 1, 2, 5, 10, 100, 1000 miles before it arrived at the earth, till at last, exceeding the limits of the Earth, it should pass into space without touching it.

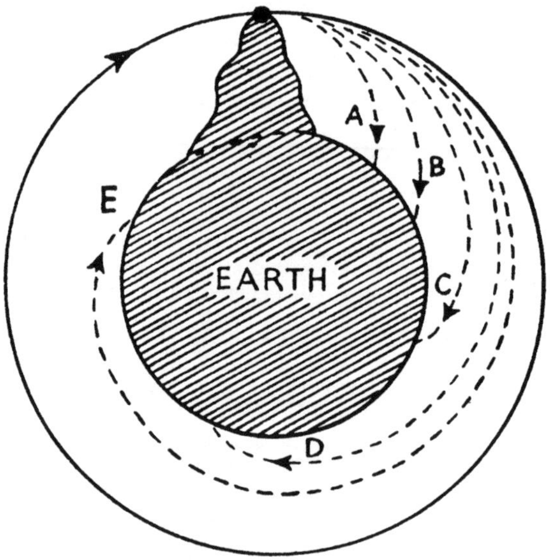

Newton's earth-satellite

Normally, a projectile (A) will soon fall to the ground. But, argued Newton, if its speed is increased sufficiently (B, C, D, E . . .) it will eventually circle the Earth completely: the Moon is just such a 'projectile', whose centrifugal tendencies are exactly balanced by its gravitational fall

Accompanying the explanation is the diagram reproduced opposite. It is clear that, even if the men of the seventeenth century had not the technical ability to construct an earth-satellite, Newton had envisaged its theoretical possibility:

If the velocity was still more and more augmented, it would reach at last quite beyond the circumference of the Earth, and return to the mountain from which it was projected. . . . Its velocity, when it returns to the mountain, will be no less that it was at first; and, retaining the same velocity, it will describe the same curve over and over, by the same law.

Provided that air-resistance can be sufficiently cut down, we can imagine terrestrial bodies projected so fast that they

will describe arcs either concentric with the Earth, or variously eccentric, and go on revolving through the heavens in those orbits just as the planets do in their orbits.

Now, the force which causes a stone to fall to the Earth is the force we call 'gravitation'. No one questioned its existence, or ever had—it was responsible, for example, for the acceleration of the ball rolling down Galileo's inclined plane. The task was to establish an exact analogy between terrestrial and celestial motion so that planetary motion, too, could be explained as 'gravitational'. Only bodies near the Earth would be governed by 'circumterrestrial' gravitation: in other cases, we should have to suppose the existence of (say) 'circumsolar' gravitation, directed towards the Sun, or 'circumjovial' gravitation directed towards Jupiter. In this way, Newton hoped to establish 'that there are centripetal forces actually directed to the bodies of the Sun, of the Earth and other planets'.

It is at this point that Kepler's discoveries start to be relevant. To begin with,

The Moon revolves about our Earth, and by radii drawn to its centre describes areas nearly proportional to the times in which they are described, as is evident from its velocity compared with its apparent [visible] diameter; for its motion is slower when its diameter is less (and therefore its distance greater), and its motion is swifter when its diameter is greater. . . . That the Earth describes about the Sun, or the Sun about the Earth, by a radius from the one to the other, areas exactly proportional to the times, is demonstrable from the apparent diameter of the Sun compared with its apparent motion.

These 'astronomical experiments', together with similar observations in the cases of Jupiter and Saturn, establish 'that there are centripetal forces actually directed (either accurately or without considerable error) to the centres of the Earth, of Jupiter, of Saturn, and of the Sun'; and the same thing presumably holds for other planets also. The basis of this proof is a mathematical theorem which Newton had worked out in the first part of his *Principia*, according to which Kepler's 'equal-area' rule was explicable in terms of a centrally-directed force.

Having established the existence of centripetal forces, he can now—in two cases at least—demonstrate that these vary inversely as the square of the distance from the centre of force. Kepler had demonstrated that the lengths of the planetary years increase as one moves out from the Sun, not in direct proportion to their distances, but in a greater proportion: e.g. multiplying the distance four times increased the planetary year eight times. (The lengths of the years, when squared, were proportional to the cubes of the distances.) The same relation applies in the case of the satellites of Jupiter: Newton now shows just how exactly it holds, by quoting the most up-to-date observations of Flamsteed and Townley. He next applies another of his mathematical theorems, extending Huygens' inverse-square calculation from simple circles to ellipses and other 'conic sections'. Kepler's Third Law could thus be explained, if one supposed that the circumsolar and circumjovial forces 'decrease inversely as the square of the distances from the centres'.

The discussion now becomes more quantitative. Newton calculates the comparative forces exerted by the Sun on each of the planets, and shows that these are probably 'proportional to the quantities of matter', i.e. the masses of the different planets. This result, he points out, is in line with his own experimental demonstration that terrestrial gravitation exerts on pendulums of different materials forces precisely proportional to their masses. (The same result explained how earlier mathematicians, such as Galileo and Descartes, could disregard the difference between weight and mass with impunity.) All gravitational forces, Newton concludes, are reciprocal: the Earth, for instance, attracts the Sun weakly while the Sun attracts it strongly. Taking the whole planetary system together, it follows that 'the common centre of gravity of the Sun and all the planets is at rest'. Only if the planets were for the moment arranged about the Sun with perfect symmetry would this 'common centre of gravity' coincide with the Sun's own centre. Normally, Newton accordingly suggests, the Sun 'moves with a very slow motion'.

At this point, Newton was conservative. The old picture of the universe as having a definite centre, which was stationary with respect to the 'fixed' stars, kept its hold on him. According to his own principles, he should have been prepared to allow, as a genuine alternative, that the planetary system as a whole was moving uniformly forward in a straight line. But, brushing aside the question what it *means* to speak of anything as 'absolutely at rest', he immediately adds, 'This is an hypothesis hardly to be admitted'. Even Newton could not embrace at once all the imaginative possibilities opened up by Bruno's vision of the infinite universe.

Newton has now explained how his extended conception of gravitation ties together the astronomical discoveries of his predecessors. But at this point there is a bonus. It always makes a theory much more convincing if, in addition to those phenomena it was introduced to explain, it also provides solutions for other, unexpected, and apparently unrelated problems. For Newton is able to demonstrate that the tides of the sea also are a consequence of gravitational interactions between the Sun, Moon and Earth. This was a most remarkable achievement: none of his predecessors had given a very coherent account of the tides. Galileo, for instance, attributed the tides to forces created by the lateral and rotatory movements of the Earth. Gravitational theory explained the tides, not just in general terms, but in detail—showing, for instance, why 'the tides are greatest about the equinoxes', and even such minor variations as the following:

But because the Sun is less distant from the Earth in winter than in summer, it comes to pass that the greatest and least tides more frequently appear before than after the vernal equinox, and more frequently after than before the autumnal.

And this was not the only bonus. In addition, Newton went on to explain the appearance and motion of comets. He supposed them to be material bodies, acted on by gravitational forces in exactly the same way as the planets, only travelling around the Sun in highly eccentric ellipses or even parabolas. Furthermore, he said, if we suppose 'that the air and vapours are extremely rare in celestial spaces', then 'a very small amount of vapour may be sufficient to explain all the phenomena of the tails of comets'. The pressure of the atmosphere drops off with extraordinary rapidity,

> In such manner, that a sphere of that air which is nearest to the Earth, of but one inch in diameter, if dilated with that rarefaction which it would have at the height of one semidiameter of the Earth, would fill all the planetary regions as far as the sphere of Saturn, and a great way beyond; and at the height of ten semidiameters of the Earth would fill up more space than is contained in the whole heavens on this side of the fixed stars.

The fact that one can actually see the stars shining right through the tails of comets shows just how thin they are.

Newton's argument had taken a course characteristic of the best scientific arguments. He had assumed at the outset a heliocentric system in empty space. His comparison of terrestrial and celestial motion had now led to the establishment of universal gravitation as a plausible general hypothesis; from this hypothesis, the tides and comets had also been explained; and finally, the initial assumptions had actually been weakened. It was not strictly necessary to assume that the Sun was absolutely at rest: in all probability, it did move slightly. Nor need the interplanetary regions be absolutely empty: there are independent reasons for supposing that they are 'vastly rarified', and this is enough. The argument had come full circle.

THE CHARACTER OF NEWTON'S ACHIEVEMENT

Only by going through Newton's argument with some care can one get a proper feeling for the *sweep* of his imagination. Logically speaking, there is nothing out-of-the-ordinary about the form of his argument. But the scale on which he works and his mathematical command are both extraordinary. Starting from facts familiar to his contemporaries and applying mechanical principles which many of them recognized—at any rate in general terms—he transformed Borelli and Hooke's 'hypothesis of attraction' into an exact mathematical generalization; and with its aid he wove the fabric of the planetary system into an entirely new pattern.

Newton's faith in the possibility of celestial physics was absolute. The right which Copernicus had claimed—to demand that mathematical astronomy must make sense in terms of physical principles—was now established past challenge. Of all Newton's contributions to science, this was probably the most significant. In themselves, the theories of lunar motion, tidal influence, and comets have been of little technological importance, at any rate before the age of the Sputniks. Even the general theory of gravitation has borne little direct practical fruit. For engineering, it is Newton's account of the laws of motion and his rules for the calculation of forces that are valuable.

From the intellectual point of view, however, the success of the theory symbolized a great deal. The ideas of astrology could at last be completely rejected. The unpredictability of comets had always made them appear ominous. Now, men understood them so well that Edmund Halley actually predicted when one would reappear. He referred to striking parallels in the records of comets observed in 1531, 1607, and 1682, and concluded that they were all appearances of the same highly eccentric planet, having a period of some seventy-five years. The comet duly returned in 1758, as he predicted—a posthumous confirmation of his prophecy.

Yet it was not any one detail of Newton's theory, but the overall conception, that was its most important feature. Taken singly, his ideas had mostly been anticipated. Buridan declared that the planets have no need of an external motive force to keep them moving; they needed such a force, as Descartes realized, only to stop them from flying off in straight lines, being kept in their near-circular orbits only if acted on by a centripetal force (analysed by Huygens) counteracting their own rectilinear tendencies. This counteracting force was probably an attraction towards the Sun, as Borelli had suggested, and might obey an inverse-square law, as Hooke and Huygens had speculated. One could even demonstrate in two lines that such an inverse-square law fitted Kepler's Third Law for the case of a circular orbit. What then was there left for Newton to contribute?

At the level of technical detail, Newton's positive additions were few: such things as a clearer analysis of the relations between weight and mass, and the demonstration that an inverse-square law of attraction would fit the case of ellipses or parabolas, as well as that of circles. If there had been no more to Newton's achievement, he would never have seemed the Prince of Scientists.

The truth lies elsewhere. Newton's unique contribution lay in the imaginative integration of many ideas into a single picture. This quality of imaginative integration is shared by many of the greatest scientific theories. Starting with a comparatively simple step, but systematically carrying the analysis through an unexpectedly wide field, such theories have the power to present old problems in an entirely new light. Whole new fields of study are opened up to patient and industrious enquiry. As a result, what had seemed to be old, insoluble difficulties appear to us in retrospect, perhaps unfairly, as mere confusions of mind.

There is, however, another aspect to all theories. However far-ranging one's imaginative speculations, they can become a part of science only if worked out with detailed proof. The full possibilities of Newton's ideas could be demonstrated only by checking its implications in complete detail. The first imaginative leaps could have been taken, and probably were, by others. But no one else at the time had Newton's wonderful mathematical capacity.

Consider, for instance, all that was involved even in his first step. He proposed to treat the Moon as a terrestrial projectile: a material body continually accelerating towards the Earth, but carried so fast sideways by its own momentum that it remained in an orbit instead of falling to the ground. This step was justifiable only if the force required to retain it in its orbit, 200,000 miles from the Earth, corresponded exactly with that which normal gravity might be expected to exert. Three problems accordingly arose—first, to compute the rate at which the Moon had to be accelerating out of a straight line, in order to move exactly as it was observed to do; second, to compare this acceleration with the acceleration of free fall at the Earth's surface, and show that these two accelerations (and so the respective forces) were in the appropriate ratio; last and most difficult, to check that the different parts of the Earth's mass could legitimately be replaced, when it came to calculating their total attractive power, by a single mass at the centre of the Earth.

The first computation was not difficult in theory: it was only laborious. The second ran up against an early snag: the estimate of the Earth's radius currently accepted in the 1660's was inaccurate, so that Newton's first calculations could be only approximate. The third problem was mathematically the most taxing: once a proof could be given that such vast bodies as the Sun and its planets behaved gravitationally no differently from concentrated point-masses, all the subsequent calculations were straightforward. But he could not just *assume* this, and finding the necessary mathematical proof held him up for a considerable time.

Matters were made slightly easier for Newton by his introduction of the 'method of fluxions', equivalent to our differential calculus. This shortened his work at the outset, but only at the cost of lengthening it later. Because his mathematical methods were novel and difficult to follow, he was obliged to reconstruct all his proofs in standard Euclidean geometry before they could be published. Even so, the scientists of the eighteenth century, though regarding themselves as Newtonians, turned with relief from the rigours of the *Principia* to the comparatively easy reading of the *Opticks*.

Newton even developed the theory of hydrodynamics—that is, the laws governing the motion of fluids—for the purpose of refuting the chief rival theory, Descartes' theory of vortices. This could be brought into line with the facts, he showed, only if one made the most implausible assumptions about the supposed universal fluid in which the vortices existed. Hydrodynamics was then, and still is, an unrewarding branch of mathematics. Descartes himself had been content to use the hypothesis of vortices as an intellectual picture, whose very vividness was its main claim on our belief. Nowhere did he carry out the sort of detailed analysis of its implications that his rival now devoted to the hypothesis.

Newton showed that two familiar facts are fatal to the theory. The first is Kepler's Third Law, that the periods of revolution of satellites vary proportionally to the 3/2th power of their distances.

> I have endeavoured in this Proposition to investigate the properties o vortices, that I might find whether the celestial phenomena can be explained by them; for the phenomenon is this, that the periodic times of the planets revolving about Jupiter are as the 3/2th power of their distances from Jupiter's centre; and the same rule obtains also among the planets that revolve about the Sun. And these rules obtain also with the greatest accuracy, as far as has been yet discovered by astronomical observation. Therefore, if those planets are carried round in vortices revolving about Jupiter and the Sun, the vortices must revolve according to that law. But here we found the periodic times of the parts of the vortex to be as the square of the distances from the centre of motion; and this ratio cannot be diminished and reduced to the 3/2th power, unless either the matter of the vortex be more fluid the farther it is from the centre, or the resistance arising from the want of lubricity in the parts of the fluid should, as the velocity with which the parts of the fluid are separated goes on increasing, be augmented with it in a greater ratio than that in which the velocity increases. But neither of these suppositions seems reasonable. . . .
>
> If, as some think, the vortices move more swiftly near the centre, then slower to a certain limit, then again swifter near the circumference, certainly neither the 3/2th power, nor any other certain and determinate power, can obtain in them. Let philosophers then see how that phenomenon of the 3/2th power can be accounted for by vortices.

Just to settle the matter, he demonstrates that a body will move in a vortex along a closed orbit for ever, only if it has one and the same density as the surrounding fluid. Few physicists have ever troubled to calculate the implications of a rival's theory with the same assiduity as Newton brought to Descartes' vortex hypothesis.

It says much for Newton's mathematical facility that he could invent one new mathematical method after another, almost incidentally, in the service of his gravitational theory; and that he engaged in the whole enterprise as though it were an intellectual pastime, while believing all the time that the questions of true importance lay in other, religious directions. The same theoretical imaginativeness, disciplined by great mathematical capacity, showed itself in his work on optics. In addition, he was a Biblical scholar, who approached his studies with some of the same care that the nineteenth-century exponents of 'higher criticism' did. Finally, he was a chemist, who had insight enough to foresee that the idea of 'attractions'—or 'binding-forces'—might be the key to an understanding of chemical change, and to guess that the forces involved were possibly of an electrical nature. (We shall meet his atomistic ideas in our next book, on matter-theory.) Yet at the end of the seventeenth century the fog still lay thick over chemistry, and Newton searched the alchemical writers in vain for a clue which would give general understanding.

By the middle of the nineteenth century, Newton had become an idol for scientists. The *Principia*, somewhat overshadowed during the eighteenth century by the *Opticks*, was reinstated for the great work it is, and came to appear almost as an incarnation of rationality itself. More recently, and especially since the work of Einstein, biographers have concentrated on the darker side of Newton's personal character—the 'fearful, cautious and suspicious temper' of which his Cambridge associates spoke—and have declared him to be 'a tortured introvert' and a 'rapt, consecrated solitary'. He was, declared Lord Keynes, 'not the first of the Age of Reason' but 'the last of the magicians'. The Book of Nature was, for Newton, written not just 'in mathematical symbols', but in the form of a cryptogram. Physical insight might enable a man to decipher this cryptogram in part, but (says Keynes) he thought that the key probably lay in the mystical tradition handed down from the Egyptians and the Chaldeans:

> For it was their way to deliver their mysteries, that is, their philosophy of things above the common way of thinking, under the veil of religious rites and hieroglyphic symbols.

Whatever we may feel about this interpretation of Newton's character and convictions, the proof of his intellect is there for all to study in the *Principia*. Edmund Halley may be pardoned for feeling that his financial investment in the *Principia* had yielded an unparalled intellectual return. He prefixed the work with a dedicatory ode. It begins as follows:

> Lo, for your gaze, the pattern of the skies!
> What balance of the mass, what reckonings
> Divine! Here ponder too the Laws which God,
> Framing the universe, set not aside
> But made the fixed foundations of his work.
> The inmost places of the heavens, now gained,
> Break into view, nor longer hidden is
> The force that turns the farthest orb. The sun
> Exalted on his throne bids all things tend
> Towards him by inclination and descent,
> Nor suffers that the courses of the stars
> Be straight, as through the boundless void they move,
> But with himself as centre speeds them on
> In motionless ellipses. Now we know
> The sharply veering ways of comets, once
> A source of dread, nor longer do we quail
> Beneath appearances of bearded stars. . . .
>
> Matters that vexed the minds of ancient seers,
> And for our learned doctors often led
> To loud and vain contention, now are seen
> In reason's light, the clouds of ignorance
> Dispelled at last by science. . . .

Halley regarded Newton's achievement—with some justice—as having an importance comparable with only four other human discoveries: the establishment of social life, the introduction of settled agriculture, the production of wine, and the invention of writing!

> Then ye who now on heavenly nectar fare
> Come celebrate with me in song the name
> Of Newton, to the Muses dear; for he
> Unlocked the hidden treasuries of Truth:
> So richly through his mind had Phoebus cast
> The radiance of his own divinity.
> Nearer the gods no mortal may approach.

In 1687, then, the outlines of a new picture of the Heavens were clearly drawn. Though some of its features would have surprised Copernicus, this picture justified his conviction that the system of the world could be understood in terms of *physics*. The Sun and its attendant planets were now to be regarded as a self-contained system of bodies moving under their mutual interaction alone, their centre of gravity being at rest by comparison with the visible stars. The stars themselves were presumably scattered for an unlimited distance beyond the range of observation.

This was a new picture, a comprehensive one, and also—recognizably—the beginning of *our own* picture. Newton's questions are our questions, his answers are the starting-point of our understanding. Much remained to be discovered: two or three additional planets within the solar system, countless nebulae beyond the visible stars. The horizon of the known universe was to expand far beyond anything which the men of 1700 could study; and, as time went on and the scale of the cosmos increased, the difficulties about space and time which Cusa had anticipated were to become very real problems. Still, for all that has happened to physics since Newton, he speaks to us in our own terms. We cannot fully understand Aristotle and Copernicus, or even Kepler and Galileo, without sympathetic intellectual reconstruction-work. To share this point of view demands an unaccustomed leap of our imagination. After 1687, following the development of astronomy demands more skill in mathematics but less imaginative insight. The language of Newton may be archaic in some of its idioms, but the fundamental conceptions he introduced have been incorporated into our twentieth-century common sense. In the present study, therefore, the year 1700 serves as a convenient destination: a watershed, so to speak, after which all the trails lead down to our own times by familiar routes. The particular ideas whose ancestry we set out to trace in this book were effectively formed by 1700; so, when we turn to the modern period, we need only map in outline the chief directions in which these ideas have developed.

Newton's *Principia* marks the watershed, not only because we still share many of his ideas, but for another more important reason. For, when we try to identify the moment at which a particular science comes of age, there is one useful test to apply. In the early stages of a science, there are commonly two separate traditions, which develop independently: a *craft* tradition and a *speculative* tradition. The crucial point to look for is the one at which these two traditions finally begin to cross-fertilize. We stand at just such a point.

Craftsmen develop and pass on techniques for manipulating, controlling or forecasting the happenings of Nature. They may be metal-workers, who develop techniques for smelting bronze and iron; or doctors, who discover how one can treat wounds and diseases; or soothsayers, who find out how to forecast ominous events such as eclipses. At the outset, these crafts are worked out on an empirical basis: men discover how to do all these things, without being able to explain how or why their techniques work. And their crafts often reach high standards before anyone can explain the reasons for their success.

By contrast there are men with speculative ideas, who want only to get behind the regular patterns in natural events to the laws and mechanisms which (they are convinced) underlie and explain them. When these men first attack their problems, no technological fruit need result: their speculations may be striking and original, yet at the outset the craftsmen may find nothing in these theories to help them. (The theories will probably not explain half of what the craftsmen are already familiar with in practice.) But eventually a time comes when the speculative tradition pays off. When this happens, the theorist's speculations can at last make the practical success of the craftsman intelligible. It becomes possible, from then on, to improve the older craft-techniques in the light of the theorist's new picture. Empirical crafts are replaced by scientific technology. To put the point in a phrase: one crucial mark of an adult science is the union of craft and speculation. By this standard, Newton's theory was the first adult system of dynamics and astronomy.

How, then, did Newton unify craft and theory? He could not at once suggest better forecasting-techniques for the planets than the computers already had at their disposal. (The only such by-product of his theory was Halley's discovery of one large comet which returned to the sky regularly every three-quarters of a century. Yet this recurrence of 'Halley's Comet' might well have been spotted empirically at any time, without reference to the theory.) In fact, for at least a century there remained noticeable discrepancies between the planetary predictions of the craft tradition and the results of the best available Newtonian calculations. These 'inequalities' provided a topic of investigation and debate right up to 1800; and even today astronomical forecasting, like tidal prediction, relies chiefly on empirically developed techniques. Newton's theories did, however, make the efficacy of astronomical forecasting *intelligible*, and this is the crux of the matter. If astronomers were in general able to predict eclipses, whereas astrologers were not such reliable prophets, this fact need no longer be mysterious. Behind the heavenly motions lay an intelligible mechanism: no such mechanism appeared to link men's fortunes with the stars.

Before we leave Newton, two last things should be said about his explanations, and about the fundamental assumptions and concepts of his theory. The important thing about his account was not the categorical forecasts it led to, but the sense it made of the traditional forecasting techniques—relating their efficacy to general hypothetical statements, in the form of 'laws of nature'. The first thing to understand is this distinction between categorical predictions and hypothetical laws.

Kepler had recognized that, as a matter of fact, the planets move in ellipses at determinate speeds. Yet, for all his theories about interplanetary forces, he could not see how to explain the elliptical character of the orbits. This fact was for him an arbitrary one, which did not hang together with the rest of his system in an intelligible way—'one more cart-load of dung [as he put it] as the price for ridding the system of a vaster amount of dung'.

Newton's theory, by contrast, showed that on certain hypotheses Kepler's discoveries would be, not arbitrary, but natural and intelligible. If it was correct to identify the planets as 'satellites moving freely under the action of an inverse-square centrally directed force only', then—on the hypothesis of gravitation—their observed motions were just what one was entitled to expect. In this way, Newton's *conditional* (or hypothetical) laws made sense of the connections lying behind Kepler's *factual* (and categorical) discoveries.

The second point to notice is this: where Aristotle's theory of motion was based on familiar, everyday principles, Newton's was stated in terms of abstract mathematical ideals. The circling heavens, a falling stone, smoke rising from a fire, the steady progress of a horse and cart: these were the objects by comparison with which Aristotle explained other kinds of motions. For Newton, on the other hand, the explanatory paradigm was a kind of motion we *never* encounter in real life. Nothing ever actually moves uniformly and free of all forces, at a steady speed and in a constant Euclidean direction. Yet Newton was able to bring together the threads left loose by his predecessors by systematically applying just this abstract ideal of 'natural' motion. So far from being guided by experience alone, he could not afford to be too much tied down to the evidence of his senses, or to the results of experiments: it was, rather, Aristotle who stuck too closely to the facts. Newton was ready to imagine something which was practically impossible and treat that as his theoretical ideal.

In his mathematics, he was equally prepared to operate with the abstract notion of 'infinitesimals'. Thus he could at last handle velocities-at-an-instant, which had eluded Zeno and Aristotle, and had been grasped only obscurely by the mediaeval mathematicians. He defined these velocities as 'fluxions', in terms of the 'ultimate ratios' or limits

> . . . towards which the ratios of quantities decreasing without limit [e.g. distances gone in given small increments of time] always converge; and to which they approach nearer than by any given difference however small, but which they never go beyond, nor in effect attain to, till the quantities are diminished *in infinitum*.

In each case, the crucial step forward from Aristotle involved a mathematical *idealization*.

How could one know that this step was justified? To begin with, it had to be taken partly on a reasonable but unprovable trust. Galileo had said, 'The book of Nature is written in mathematical symbols'; and now Newton seemed to have deciphered some of the code. Yet there was no way of showing at once that his reading of the clues was definitely correct; nor was it certain that idealization was the right procedure. (There can be no set recipes for theory-making.) In any particular scientific situation, one can only back one's reasoned hunches with a lot of hard work and thought. The fruit of one's speculations (if there is going to be any) will come later.

FURTHER READING AND REFERENCES

The standard biography of Newton is by L. T. More. There are also brief lives by E. N. da C. Andrade, J. W. N. Sullivan and S. Brodetsky. The only current edition of Newton's *Principia* is that of F. Cajori; but a full scholarly edition is in preparation by I. B. Cohen and A. Koyré. For the general intellectual background to Newton, consult

Basil Willey: *The Seventeenth Century Background*
E. A. Burtt: *Metaphysical Foundations of Modern Science*
E. W. Strong: *Metaphysics and Procedures*
A. Wolf (and D. McKie): *A History of Science Technology and Philosophy in the sixteenth and seventeenth Centuries*
H. T. Pledge: *Science since 1500*

See also the discussions in Butterfield and Hall, already referred to.

There is an admirable selection of Newton's non-mathematical writings, in the book

H. S. Thayer (editor): *Newton's Philosophy of Nature*

10

The Widening Horizon

A WATERSHED marks the end of an uphill grade, but is not the end of the road. Newton's fresh ideas immediately made sense of a wide range of natural happenings; and they were in due course to become part of the fabric of everyone's 'common sense'. But they still had to develop further, and problems which subsequently arose forced men to have second thoughts about the basic assumptions of the system. The next major intellectual step in astronomy and dynamics dealt with precisely these problems—and resulted in the introduction of Einstein's two theories of relativity.

Newton's work was meanwhile having its effects outside physics also. The intellectual confidence which his success created spread to other sciences: chemists in the eighteenth century made good use of his idea of the conservation of mass, while psychologists and social scientists borrowed his language and tried to discover mental and economic 'forces' to explain the happenings in their own fields. Even in literature and theology, the new confidence left its mark. Instead of treating cosmology as a matter for revelation rather than science, Protestant theologians began to welcome the advance of science as an aid to 'natural theology': seeing in the orderliness and arrangement of the natural world evidence of God's nature and activity. This phase was to pass, but for nearly a century many theologians regarded Newton's work—quite literally—as a God-send.

THE LOOSE ENDS: (1) PLANETARY INEQUALITIES

What Newton's theory succeeded in explaining, it explained very well. But it had not answered all men's questions about the solar system: for instance, why the different planets were set at their respective distances from the Sun. For a time astronomers continued to believe that some simple mathematical law might relate these distances; but by now we no longer expect to find any such tidy law. Others of these questions had, however, to be dealt with in due course. Newton himself was struck by the fact that his theory explained only the *continued* motion of the planets along their tracks. There was nothing in it to show how they got into this regular and stable system in the first place:

> Where there is no air to resist their motions, all bodies will move with the greatest freedom; and the planets and comets will constantly pursue their revolutions in orbits given in kind and position, according to the laws above explained; but though these bodies may, indeed, continue in their orbits by the mere laws of gravity, yet they could by no means have at first derived the regular positions of the orbits themselves from those laws.

In Newton's personal opinion, the origin of the solar system must be the deliberate design of the Creator:

> This most beautiful system of the sun, planets and comets could only proceed from the counsel and dominion of an intelligent and powerful Being. And if the fixed stars are the centres of other like systems, these, being formed by the like wise counsel, must all be subject to the dominion of One, especially since the light of the fixed stars is of the same nature with the light of the sun and from every system light passes into all the other systems; and lest the systems of the fixed stars should, by their gravity, fall on each other, He has placed those systems at immense distances from one another.

Yet there is a difficulty about any 'natural theology' which invokes the actions of the Deity to account for regularities not already explained by science. Men were bound, before long, to try to explain the origin of the planetary system in non-theological terms, just as they had previously looked for physical theories to explain its continued regularity. By treating everything not yet scientifically understood as miraculous and Divine, one puts theology in a position from which it will be obliged to retreat whenever knowledge advances. So, in this case, Kant and Laplace in due course put forward a *physical* hypothesis, called the 'nebular' hypothesis, about the formation of the solar system. On the whole, it was preferable to think of God as creating and arranging the laws of the physical world in the beginning in such a way that He did not subsequently need to intervene in its affairs—at any rate, so long as the actions of human beings were not concerned.

Newton himself detected the hand of the Almighty in the order of the solar system. And at this point he ran into a difficulty. As he believed, God had created it according to an ingenious plan, so that it would continue to move in an orderly and stable manner for ever. Yet nothing in his theory guaranteed this: on the whole the evidence pointed the other way. The difficulty was, in part, mathematical. Suppose you consider a system of two bodies only—for instance, the Sun and any one planet. You can then show mathematically that the orbit resulting from their gravitational interaction alone will be stable. Yet the planetary system comprises at least twelve bodies, which interact not only with the Sun but also (though more weakly) with each other. In consequence, each of the planets has slight 'perturbations' in its orbit: it does not move exactly as it would do if the other planets were absent. Now, the question arises: could these perturbations become cumulative, so that a planet was eventually edged right off-course? Or, taking everything together, do they tend to cancel out and leave the whole system of planets stable? As a matter of mathematics, the relevant equations in Newton's theory can be given no general solution: the question can be answered only step by step, by 'successive approximations' calculated afresh for each particular configuration of planets.

Furthermore, as Newton conceded, there are many mechanical transactions going on in the world which have the effect of dissipating the total amount of motion in the universe:

> By reason of the tenacity of fluids and attrition of their parts, and the weakness of elasticity in solids, motion is much more apt to be lost than got and is always upon the decay. For bodies which are either absolutely hard or so soft as to be void of elasticity will not rebound from one another.

It therefore seemed possible that God had to intervene in the natural world so as to redress the disorder introduced by the 'perturbations', and restore the dissipated motions. For this purpose, He must utilize certain 'active principles'.

> Seeing, therefore, the variety of motion which we find in the world is always decreasing, there is a necessity of conserving and recruiting it by active principles, such as are the cause of gravity, by which planets and comets keep their motions in their orbs and bodies acquire great motion in falling, and the cause of fermentation, by which the heart and blood of animals are kept in perpetual motion and heat, the inward parts of the earth are constantly warmed and in some places grow very hot, bodies burn and shine, mountains take fire, the caverns of the earth are blown up, and the sun continues violently hot and lucid and warms all things by his light. For we meet with very little motion in the world besides what is owing to these active principles. And if it were not for these principles the bodies of the earth, planets, comets, sun and all things in them would grow cold and freeze, and become inactive masses; and all putrefaction, generation, vegetation, and life would cease, and the planets and comets would not remain in their orbs.

For the German philosopher Leibniz, these aspects of Newton's theory were so repugnant that he felt justified in sweeping the whole system aside. As he saw it, Newton had begun by claiming that the wisdom and dominion of the Almighty were evident in the design of His celestial mechanism, but ended by crediting Him with a botched job—a cosmos which required to be 'adjusted' and 'wound up again' from time to time. Was God then so poor a craftsman, Leibniz asked, that He could not make a world which ran accurately without His continual intervention?

The problem of the conservation of motion was largely cleared up in the eighteenth century, when d'Alembert stated clearly for the first time the relation between 'momentum' and 'kinetic energy'. The problem of the stability of the solar system has not been wholly solved even to this day. Since the 'three-body problem' can be dealt with only approximately, all one can show is that this stability appears at the present time to be assured to a high order of accuracy: there is no guarantee that it must be assured for ever. However, after a time, the question ceased to be urgent, for two reasons. Natural theology lost some of its naïve charm. Men ceased to think of God as a 'craftsman' in such a literal manner as Newton and the early eighteenth-century divines had done. And further, the actual magnitude of the orbital perturbations turned out to be much less than had been supposed.

At the end of the eighteenth century, the French mathematician Laplace, using more refined methods and data, recalculated the Newtonian orbits of the Moon and the planets: he showed, as a result, that the observed motions were almost exactly those required by a strict application of the theory. So he was able to argue that no Divine intervention was needed to keep the planetary system in order: when he was asked why he had eliminated the action of God from the theory of the celestial mechanism, he replied, 'I had no need of that hypothesis'. This story in no way implied that Laplace was an atheist. On the contrary, he was a religious man, but one with a more orthodox theology than Newton's had been. For Laplace, it was a mistake to treat the action of God in the natural world as a species of explanatory 'hypothesis'. God was rather a 'transcendent' being, who should not be expected to manifest Himself directly in the operations of physical Nature.

Even after Laplace the coincidence between calculation and observation was still not perfect. In 1781 that industrious observer William Herschel had discovered the planet Uranus. As the motion of this new planet was progressively tracked year by year, mathematical astronomers attempted to determine what precise ellipse it was following—allowing, of course, for the perturbations introduced by its proximity to Saturn. Yet, whatever size and eccentricity they assumed the orbit to have, the planet continued to depart slightly from their predictions: by the 1840's, Uranus had been followed around nearly one complete circuit, and the anomalies in its motion had become a minor astronomical scandal. Leverrier and Couch Adams calculated independently that the unexplained deviations of Uranus might, on the Newtonian theory, be the consequences of another unknown, and more distant, planet perturbing its motion; and, by a mixture of good judgement and lucky guesswork, they worked out where in the sky it might be found. Thus Neptune was discovered. (In the twentieth century one further planet, Pluto, has been located; though there is some evidence that it may originally have been, not a planet, but a satellite of Neptune, which subsequently 'escaped'.)

At the time, Leverrier's prediction of the existence and position of Neptune appeared to represent the Newtonian theory's unanswerable triumph. After 150 years of hesitation on account of the planetary inequalities, men saw in this discovery a splendid confirmation of its validity. For, had Newton's conclusions not been correct (it seemed), Leverrier's success would surely have been a fantastic coincidence. Yet there was one snag about this argument. Uranus was not the only planet whose motions still showed inequalities. Mercury, too, behaved in a slightly irregular way: the axis of its ellipse was slowly changing its direction relative to the Sun. It was doing so at a minute rate (less than one degree a century), but still, faster than could be explained. Leverrier next tackled this problem. To explain this abnormal behaviour of Mercury, he argued precisely as before, and so postulated a small planet, situated between Mercury and the Sun: this he named Vulcan. For some years observers kept reporting that they had sighted Vulcan, but probably they had seen only sunspots.

Eventually it was accepted that Vulcan does not in fact exist. The moment of triumph for the Newtonian theory had been, also, the moment at which it displayed its crucial weakness. After searching in vain for any alternative explanation, astronomers were forced to assume that the force of gravitation did not vary with *absolute* exactitude as the inverse-square of the distance: in the case of extremely massive bodies acting on each other at comparatively short distances, the force must be slightly *greater*.

At the outset this assumption was an arbitrary one. Then, during the First World War, Albert Einstein put forward his general theory of relativity. This involved a slight modification of the original Newtonian Law. The change was a small one, whose consequences could show up only under extreme conditions: yet it was of just the right order of magnitude to explain the anomalies in Mercury's motion. It also implied that light-rays would be gravitationally deflected in the neighbourhood of heavy bodies more than had previously been foreseen. Immediately after World War I, an attempt was made to detect this latter effect by photographing stars near the Sun's disc during a total eclipse, and the results seemed at the time to favour Einstein's view. Yet the effect was minute, and measurements of extreme accuracy were needed to choose between the new view and the old. So it remains a matter of controversy how far the 1921 observations really bore out Einstein's general theory.

THE LOOSE ENDS: (2) THE MECHANISM OF GRAVITY

These doubts about the Newtonian theory were reinforced by other considerations. This second dispute turned on the central notion of Newton's theory, that of 'gravitational force'. Here again, Newton had immediately been attacked by Leibniz. (This was quite apart from the violent controversy between their followers, as to which of the two men had been first to invent the differential calculus.) Leibniz agreed with Newton in thinking that the natural motion for a body free of external forces was straight-line motion. But the only kind of 'external force' he would admit was one transmitted by direct contact on impact: e.g. the force of a tennis racquet on a ball. As a result, he could not stomach Newton's apparent belief that the 'force' of gravitation could act across empty space, between the Sun and a planet 100,000,000 miles away. If Newton had explained the mechanism by which gravitational action was transmitted, Leibniz might have taken his theory more seriously: as things stood, he found the idea entirely unacceptable. It made of gravity (he said) a perpetual miracle.

Leibniz's difficulty was widely shared, and Newton was very sensitive on the point. In his reply, he insisted that his theory had been intended to demonstrate the action of gravity only 'in a mathematical way': i.e. to establish, not its cause or mechanism, but only the form of law manifested in its effects. Both he and Samuel Clarke (who carried his banner in the resulting controversy) protested, as loudly as Leibniz, that some kind of push-and-bang mechanism must eventually be discovered for gravity. Certainly—Newton expostulated, in a letter to Richard Bentley, the Bishop of Worcester, who had been his colleague at Trinity College, Cambridge—gravity was not in his view an 'occult property', like the 'virtus dormitiva' in opium at which Molière poked fun:

> It is inconceivable that inanimate brute matter should without the mediation of something else which is not material, operate upon and affect other matter without mutual contact, as it must be if gravitation, in the sense of Epicurus, be essential and inherent in it. That gravity should be innate, inherent, and essential to matter, so that one body may act upon another at a distance through a *vacuum*, without the mediation of anything else, by and through which their action and force may be conveyed from one to another, is to me so great an absurdity that I believe no man who has in philosophical matters a competent facility for thinking can ever fall into it. Gravity must be caused by an agent acting constantly according to certain laws, but whether this agent be material or immaterial I have left to the consideration of my readers.

Still, for all his protestations, Newton had in fact done nothing to explain the underlying mechanism of gravitational action. He speculated about it in some of his letters, but declined to put forward any hypothesis in his published works, saying:

> And to us it is enough that gravity does really exist, and acts according to the laws which we have explained, and abundantly serves to account for all the motions of the celestial bodies, and of our sea.

This problem, too, had to be left for later generations, and it has not yet been solved to everybody's satisfaction. Newton himself suspected that gravitational influence was transmitted through an 'ether' made up of imponderable atoms, which acted on one another by direct impact and pressure; but this idea has long since been given up. For a while it seemed that gravitation might be a species of radiation, and this idea is still revived from time to time. Immanuel Kant, for example, treated gravitational force as an influence propagated uniformly through space in all directions; he argued that, for reasons of geometrical symmetry, the force of gravity *must* vary as the inverse-square distance. This parallel between gravitation and radiation is kept alive today in the suggestion that gravity-waves are affected by the 'wave-particle duality', having particle-like 'gravitons' as the counterpart to 'photons' of light.

By 1900 the problem was—notoriously—as unsolved as it had been in 1700. Moving bodies unquestionably deviate from Euclidean straight lines whenever they pass near other massive bodies, in close conformity to the mathematical laws that Newton had presented. Yet if one accounted for this fact by saying (for instance) that the Sun exerts a 'force' on the Earth, no mechanism could be pointed out to explain this interaction. This meant accepting the bald fact that the Sun—way over there!—could exert a force on the Earth—just here under our feet!—without any rope joining them, or any jostling atoms or beam of radiation transmitting its attractive forces. And this was quite a lot to have to accept as a bald fact. After all, the empty spaces between the Sun and the Earth act as an acoustic shield, preventing ordinary *sound* from being transmitted across the gap. So how was it that gravity was not shielded in the same way, but acted across an effective vacuum?

Some physicists considered that this problem of action-at-a-distance could be got around by introducing the idea of 'fields of force', and this form of account has been popular. A 'field-strength' of such-and-such at any point would then explain (they argued) why a body moving in the field was accelerated from a straight line at that point by just the corresponding amount. Other people were not sure that the term 'field' helped one very much. If the action of this field could itself be accounted for as, for example, the result of a state of strain in an elastic medium, all well and good: an explanatory mechanism would have been produced. Otherwise the new term, instead of overcoming the paradoxes of action-at-a-distance, served only to inure us to them. Leibniz would have said that the term 'field' was only a way of making Newton's miracles computable. Failing the discovery of an ethereal medium to transmit gravitational forces (and this was never revealed by *experiment*), the nature of gravitational action remained a mystery.

A more profound response to the paradoxes was to question the seventeenth-century conception of a 'mechanism', as necessarily being a push-and-bang affair. Even if such a mechanism could be found, in the case of gravity, would it provide a satisfactory solution of the problem? Suppose there *were* a rope joining the Sun to the Earth: that would not fundamentally help us. It would only shift the problem one stage back—since it arises again as soon as you try to explain why each inch of the rope coheres to the adjoining inches on either side, instead of coming apart from them. That, too, can be accounted for only by supposing some internal field of binding-forces which act at a distance, from atom to atom, and prevent the rope from falling into pieces. (Newton himself had shrewdly noted that the cohesion of solid bodies poses a problem, and should not be taken entirely for granted.) One way or another, then, the existence of force-fields had begun, by 1900, to look like one of those basic principles of Nature on which all explanation must depend: an ultimate explanatory category such as men had been seeking ever since 600 B.C. Yet one further suggestion must be mentioned before we leave 'universal gravity'; since this topic, too, has served as one starting-point for Einstein's theory of relativity.

To see the nature of this suggestion, we must review the argument which led Newton to postulate a 'force' of gravity. He had taken over from Bruno and Descartes the idea that Euclidean straight-line motion was the natural ideal of free motion, referring (for instance) to 'the rectilinear courses which, left to themselves, the planets should have pursued'. Comparing the actual motions of the planets with this ideal, he was at once obliged to suppose the existence of a 'gravitational force': 'From the laws of motion, it is most certain that these [gravitational] effects must proceed from the action of some force or other.' Yet this new force was, in more than one respect, untypical of forces-in-general. Quite apart from the absence of any mechanism of transmission and the impossibility of shielding against its action, it lacked the sensible effects that we associate with normal forces. When you jump off a wall, you *feel* no force pressing you down—your body simply accelerates.

Only Newton's insistence on the rectilinear ideal had compelled him, accordingly, to introduce this new 'force'. Yet what if this ideal had, in fact, been *too* abstract? This paradigm of free motion was, after all, part of Newton's theoretical scaffolding, rather than the authentic voice of Nature herself: so might it not be better to modify that paradigm, instead of introducing a hypothetical force, and *avoid* the problem of the mechanism of gravitation instead of solving it? Newton had rejected Aristotle's definition of 'natural motion', because it involved references to the Earth as being 'the Centre of the Universe': in a Copernican picture, the Earth could not be given a preferential status as compared with any other astronomical object. Yet Descartes' straight-line ideal of natural motion was not the only alternative to the Aristotelian view. W. K. Clifford had suggested, some years before Einstein, that all massive bodies alike might be the centres of regions in which unforced motion was *naturally* non-rectilinear, and Einstein took this idea up.

He had one additional reason for doing so. As he saw, one further assumption underlay Newton's theory. Spatial and temporal magnitudes were there defined in a way quite independent of the properties of actual material objects and the forces acting between them. Yet might not a large mass at one position in space affect our *measurements* of these magnitudes in its vicinity? Could we be *sure*, for instance, that clocks ran at exactly their usual rate when near centres of intense gravitation? In fact, argued Einstein, the distinctions that Newton had drawn between 'space', 'time', 'matter', and 'force' were both too abstract and too sharp. Spatio-temporal measurements have to be made using physical agencies—ideally, light-signals: and it is no use talking about two events as (say) 'simultaneous', unless some conceivable method exists for checking this simultaneity.

When this fact is allowed for in our theoretical analysis, we get a structure of theory different from Newton's: one in which the critical distinctions cannot be drawn so sharply. In such a relativistic picture, one body 'gravitates' towards another along a 'natural' path and no 'force' comparable with the forces of impact need come into the account. Thus the presence of matter once again affects the geometrical shape of a body's natural track, as it had earlier done in Aristotle's theory—but with one crucial difference. The Earth is now a centre of gravitational motion, not because it is the centre of the universe, but simply because it is a large, massive body; and, in the new picture, *all* such bodies are centres of gravitation.

At the same time that the loose ends in the Newtonian theory were being tidied up, the frontiers of the observable universe were being pushed back. Newton himself spoke vaguely of the visible stars as 'fixed', and nothing in his system indicated what lay beyond them. As time passed, men came to take it for granted that they were at different distances from the Earth, even although no direct estimates were available. The phenomenon of stellar parallax, whose existence had been crucial for the heliocentric theory since Aristarchos, was finally established beyond doubt only in 1837. Bessel in Germany and Struve in Russia then showed that earlier claims to observe the effect had been unfounded, but that it did authentically appear in the case of the stars 61 Cygni and Alpha Lyrae: in each case, the angular displacement produced by the Earth's seasonal motion was less than one ten-thousandth of a degree. Sir John Herschel, in presenting the Royal Astronomical Society's medal to Bessel, spoke with understandable enthusiasm:

> I congratulate you and myself that we have lived to see the great and hitherto impassable barrier to our excursions into the sidereal universe . . . against which we have chafed so long and so vainly . . . overleaped.

Towards the end of the eighteenth century the question of the origin of the planetary system was raised again. It had long been known that there were certain 'nebulous' stars, which through a telescope gave the appearance of a whirling mass of luminous gas: Galileo had remarked on them in a passage quoted earlier in this book. Kant and Laplace now independently suggested that the solar system had originated in such a whirling nebula, which had condensed and divided into the Sun, planets, and satellites. On studying the nebulae further, men recognized that many of them were in fact far larger than the solar system. Rather, they suggested, the solar system itself was only a small part of a much vaster nebula. The Milky Way, or 'galactic' band, might then be just our transverse view of our own total 'galaxy'. (Plate 14.) And this system might be only one of many similar ones. Beyond our galaxy, it appeared, lay other similar ones, at enormously remote distances. This was the beginning of an expansion in the astronomical horizon far beyond that which Newton himself had known.

During the last half-century, some cosmologists have even asked whether we may not be approaching the very boundaries of the universe itself. At this point, the difficulties raised by Nicolas of Cusa become active again. In any literal-minded sense, questions about 'the Boundary of Space' can have no meaning, and the same is true of questions about 'the Beginning of Time'. In his youth, Kant himself was an enthusiastic astronomer and physical cosmologist, and believed that he could explain the whole history of the universe since the Creation on strict Newtonian principles. But in his later years he saw that this was asking too much of physical theory. Questions about 'the Whole Universe' (as they were then understood) landed one in the very paradoxes that Cusa had foreseen. In a relativistic picture of the universe, by contrast, some of the paradoxes can be avoided: for instance, one can imagine the *dimensions* of the universe being finite, without on that account being forced to suppose the existence of a paradoxical 'boundary'. And, in recent years, certain new phenomena have caused the whole question to be reopened. Several exciting new 'world-models' have been proposed, to account for the observed distribution of the heavenly bodies.

The discovery which has done most to stimulate speculation has to do with the optical spectra of the distant galaxies. At this point, a little background is required. Aristotle had believed that celestial bodies were made of a different stuff from earthly ones: the unchanging 'quintessence'. Philoponos disputed this, arguing that they gave off light similar to terrestrial light, and so were probably ordinary material bodies. But it was only with the development of prisms and spectrum analysis that men could at last identify the actual substances composing distant Heavenly bodies. By comparing the spectrum lines visible in the light emitted by the Sun, for example, with those given off by known substances on the Earth, Fraunhofer and his successors demonstrated the presence of sodium and other terrestrial substances in the outer layers of the Sun's gaseous sphere. Eventually the argument was even used in reverse. Sir William

Ramsay identified certain unfamiliar spectrum lines in the Sun's light as belonging to an unknown substance, which he named 'helium', and later went on to prove that it existed in minute quantities in the Earth's atmosphere also.

With this record of striking success in mind, astronomers assumed until recently that celestial and terrestrial spectra were identical. But Hubble has shown that this is not absolutely the case: light from the more distant galaxies reaches our telescopes with all its spectrum lines shifted slightly towards the red end of the spectrum —i.e. their 'wave-length' is increased. The further away a galaxy is, the greater is this 'red-shift', its amount seemingly increasing in proportion to the distance. The crucial question for physical cosmology today is: how is this red-shift to be interpreted? On the one hand, it can be put down to a recession of the nebulae: the size of the shift varying with the speed of the recession. On this account, the most distant galaxies would have to be moving apart at enormous speeds, and some cosmologists believe that we are living in the aftermath of a 'cosmic explosion'. They see the receding galaxies as like the fragments of a gigantic hand-grenade or 'primeval atom', which burst apart from each other five thousand million years ago and are still scattering.

Alternatively, it has been suggested that the red-shift is not evidence of recession at all. Let us suppose, for instance, that radiation gradually 'ages', its wave-length increasing slowly as time goes on. In that case, even if the galaxies were all stationary, the light which had come the greatest distance and so been longest in transit would arrive at the Earth with the greatest red-shift.

These are not the only possibilities. Even supposing the recession to be genuine, theory can still proceed in several directions. Some cosmologists reject the 'primeval atom' account: they argue rather, that the Universe is, on balance, in a 'steady state'. It looks on an average the same to observers at all places and all times, and the continual recession of matter beyond the range of effective observation is counteracted by the continual coming-into-existence of hydrogen atoms out of nothing. (This theory would have the merit of explaining why, in the observable part of the universe, hydrogen is by far the most prolific of the chemical elements.)

Physical cosmology today is in a curious logical condition. The questions are still largely obscure, the evidence is scanty, the inferences are tenuous, and the scales are evenly weighted between the various theories. On the other hand, the intellectual stakes are so high that men naturally—and very properly—find the subject exhilarating. If it did in fact prove possible to establish, by rational enquiry, any convincing theory about the birth (or rebirth) of the material world we now know, that would be an intellectual triumph as exciting in its own way as anything which Isaac Newton achieved.

THE WIDER INFLUENCES OF NEWTON

So far, our subject has been Newton's influence in his special fields: dynamics and astronomy. What Euclid had done for plane geometry, and men from Heytesbury to Galileo had done for kinematics, Newton did for dynamics: he developed at last a systematic and fruitful account of 'force' and 'mass', and applied this to astronomy. But, once established, these ideas were of potential value in other sciences also, and they soon found application elsewhere.

In the next book we shall find Lavoisier using the conservation of mass as an axiom of his chemical theory, and similar extensions were soon made of the idea of force. Electrified or magnetized bodies accelerating towards each other (men argued) were moving under electrical and magnetic forces, just as the planets were doing under gravitational forces: and again the sizes of these forces could be estimated from the accelerations they produced in bodies of different masses. In this way the idea of force became as crucial for an understanding of electric and magnetic interactions as it had been earlier in the theories of impact and gravitation. Einstein's later years were, in fact, spent in the attempt to frame a unified theory in which electrical and magnetic forces were to be integrated into a relativistic framework, as gravitational ones had been earlier.

Newton's influence was not restricted, however, to the ideas which were his chief monument. Given his personal prestige, even his *supposed* beliefs were regarded by some people as 'the Law and the Prophets', against which no appeal could be made. For instance, he had advocated an atomic theory—or, as it was then called, a 'corpuscular' philosophy—of matter, light and the ether. This atomic theory had no serious connection with his dynamical and gravitational ideas, and he advanced only the slightest amount of evidence in its support. Yet throughout the eighteenth century Newtonian scientists accepted these corpuscular views as the Gospel truth. The results were sometimes unfortunate. Newton had been careful not to commit himself to any final theory about the physical nature of light. In discussing optical phenomena, he treated light as corpuscular, certainly; but he also attributed to it a 'periodic' or wave-like character—what he called 'alternate fits of easy reflection and refraction'—in order to account for the diffraction effects still called 'Newton's rings'. His followers, by contrast, passed over this wave-like aspect of his theories, and treated the corpuscular theory of light as having complete authority. As a result, the wave-theory of light, which might have grown naturally out of Newton's views on 'fits', was given a serious hearing only after 1800.

Most influential of all was the example Newton set to men working in other disciplines. In intellectual studies as elsewhere, it is easy to be disheartened by obstacles. The deference mediaeval scholars showed to theological authority originated in something more than subservience. They were quite genuinely modest and unassuming: they lacked the philosopher's burning conviction that disinterested enquiry can lay bare the workings of Nature. Newton's spectacular success changed men's attitudes. Doubts about the capacities of the human reason were swept away—or at any rate became unfashionable. Jonathan Swift might still poke fun at the Royal Society in the year before Newton's death: the 'Voyage to Laputa' in *Gulliver's Travels* is a sour caricature of his scientific contemporaries. Yet, before long, David Hume—to mention only one name—was claiming to have introduced the methods of Newton into the human sciences, and was writing essays on anthropological and social subjects of a kind that foreshadowed modern economics and sociology. By the mid-1700's the foundations of political and social affairs were being discussed in France by men who wished to re-order them on a 'scientific' basis. Laplace was not only a mathematical physicist: he was also a social scientist, and wrote a treatise on probability. This discussed, among other things, the mathematical techniques for determining the decisions of political assemblies.

There remains Newton's influence on religious and theological ideas. In Newton's own eyes, Nature was the miraculous contrivance of a supremely-ingenious Creator, maintained in order by His intervention:

> Blondell tells us somewhere in his book of Bombs that Plato affirms that the motion of the planets is such as if they had all of them been created by God in some region very remote from our system and let fall from thence towards the Sun, and as soon as they arrived at their several orbs their motion turned aside into a transverse one.... So, then, gravity may put the planets into motion, but without the Divine Power it could never put them into such a circulating motion as they have about the Sun; and therefore, for this as well as for other reasons I am compelled to describe the frame of this system to an intelligent Agent.

Newton's great continental advocates, Voltaire and his successors, substituted for this a very different picture of the universe—a giant, self-contained machine, set going (perhaps) by God at the Creation of all things, but left to churn on ever since without further Divine interference, according to inexorable laws. For the Deists of the eighteenth century, the Creation was God's first *and only* intervention in the affairs of His natural creation. Any other attitude smacked of superstition.

Newton's theory had removed the last justification for regarding comets with awe or alarm: this was now treated as the prime example of irrational fear. A disastrous earthquake at Lisbon provoked the fiercest debate of all, and divided the men of eighteenth-century Europe into sharply opposed camps. Some could still interpret the event as a Divine visitation—apparently ignoring the remarks of Jesus about 'the men on whom the Tower of Siloam fell'. God, in their view, still intervened through the agency of Nature to reward the good or punish the wicked: a similar idea recurred in Britain in 1940, at the time of 'the miracle of Dunkirk'. The rationalists and the Deists of the eighteenth century retorted that Divine intervention of this kind was completely out of the question. Voltaire equally ridiculed Leibniz's view that the need for a 'sufficient reason' in God's created world guaranteed that 'everything was for the best in the best of all possible worlds'. On the contrary, he argued, natural events of all sorts happen in a law-governed way, and can have no moral significance whatever, either for good or for ill. In his story *Candide*, Voltaire obliges Leibniz's spokesman to live through the Lisbon earthquake, and then mocks at him for trying to explain away the consequent sufferings as having been really 'all for the best'.

During the nineteenth and twentieth centuries, the points of contact between science and theology have changed. As late as 1650 poets could still write about the Heaven of religious faith in quasi-astronomical terms:

> My soul, there is a Country
> Far beyond the Stars,
> Where stands a wingèd Sentry,
> All skilful in the Wars.

Just so long as men could not actually find out anything about the remoter depths of the sky, this traditional imagery could retain its hold on their minds. After Galileo it was doomed. Scholars had long felt that a literal-minded conception of Heaven as lying beyond the boundaries of the natural world had grave defects: after 1610, this conception could no longer keep its place even as a visually attractive figure of religious thought. Copernicus, Galileo and Newton might not have answered all the questions science could ask about the remoter heavens, but they did remove them from the sphere of impenetrable mystery.

In the last 250 years, everything has tended to reinforce the belief that the universe is fairly homogeneous: the remoter heavens and the nearer heavens being, scientifically speaking, largely similar. Heaven, in the religious sense of the word, is accordingly no longer a topic for astronomy, and has been eliminated from celestial maps. Astronomy and theology have not parted company utterly. Problems of physical cosmology, for instance, still appear to some as having a potential significance for theology. Aquinas found it hard to reconcile Aristotle's thesis that the universe had always existed with the Christian doctrine of the Creation: so, too, for similar theological reasons, some people nowadays prefer the 'primeval atom' theory of the origin of the universe to the 'steady-state' theory. This, however, is the single point of contact remaining. Otherwise one is confronted with a striking change: the more that has been found out about the heavens from a scientific standpoint, the less significant they have proved to be from the theological point of view. Now in the twentieth century, when our astronomical knowledge is more extensive and detailed than ever before, theologians also find in astronomy less material of relevance to their problems than in any previous age.

CERTAINTY AND SCIENTIFIC THEORY

During much of the eighteenth century men saw Newton, not just as having established provisional laws for calculating the motions of bodies, but as having actually revealed the true laws for good and all. His theories seemed, that is to say, not only valid up to a point, but correct, certain and final. Later, logical doubts and perplexities began to arise. One question was particularly perplexing. Physics was an empircal science, concerned with the actual workings of Nature; Newton's evidence was localized and restricted to a limited period of time; how, then, could he have hit on laws and principles holding good for all time and all places? When the Ionians began their own search for the 'eternal principles' of things, Heraclitus had raised a similar question. Seeing that all the evidence of our senses is in flux (he argued), we can never hope to achieve certain knowledge of such 'hidden principles'. Even as late as Osiander, it had been possible to argue that Divine revelation alone could give certainty. Yet now—or so men thought at the time—Newton had succeeded in the impossible.

Consider, for instance, Newton's laws of motion. If one believes that the world was built to operate according to the specification of a Divine Artificer, then perhaps God might be supposed, as an act of Grace, to let diligent men discover for themselves what His specification had been. Yet even this answer is not wholly satisfactory: it is surely as difficult to prove for certain that an act of Grace has in fact taken place, as it is to establish the unchanging laws of Nature directly. As the eighteenth century went on, men began to ask whether the certainty and timelessness of Newton's dynamical theory must not come from some alternative source.

Once again Kant's changing views are a useful index. When young, he took it for granted that the laws of motion and gravitation provided the means of discovering what he called *The General Natural History and Theory of the Heavens.* He trusted these laws (that is to say) to provide positive information about the actual course of events by which the universe has developed. Yet Newton, in his theory, had argued about how bodies *must* move, not just about how they *do in fact* move; and the more Kant thought about this matter the less he could reconcile this necessity with the informative character he had attributed to the laws of motion. Take the first law of motion—that 'a body will persist in a state of rest or uniform rectilinear motion, unless acted on by an outside force'. This could either be a straightforward statement about what always *does* happen, or a statement about what *must* happen: it surely cannot be both. If this axiom did involve any sort of 'necessity', we could hardly find its origin in the facts of Nature alone.

For a time, it was suggested that the origin of this necessity was linguistic—lying in our definitions. If a body *must* move uniformly, unless acted on by a force, that was a consequence of the way in which we define 'force'—we use the ideal of uniform motion as our test for recognizing when a 'force' is acting. Yet, in Kant's views, this could not be the whole story. It was one thing to say, 'Unless a body is black, white, grey or transparent it *must* be coloured, because—by our own usage—that is what the word "coloured" is defined to mean': but in the case of the term 'force' the corresponding doctrine appeared unconvincing. In the axioms of mathematical physics, and in certain fundamental principles of philosophy, Kant found a kind of certainty which was neither the contingent certainty attaching to well-attested facts—nor the trivial certainty attaching to truths of definition. In order to explain the source of this 'synthetic necessary truth', he embarked on the enquiry which resulted in his great but obscure treatise, the *Critique of Pure Reason.*

For our purposes the actual conclusions of Kant's enquiry are beside the point. Nor need we any longer accept his belief that Euclidean geometry and Newtonian dynamics are uniquely applicable to the world of Nature. Yet there is one thing in his analysis which is of permanent importance. Certain sorts of principle, he argues, are the indispensable framework for any rational picture of the world. We can define 'coloured' bodies so as to include black, white, and grey ones, or alternatively so as to exclude them, entirely as we please. The decision is an arbitrary one; and to say, 'A body *must* be coloured unless it is black, white, grey or transparent' is trivial, just because our definition is arbitrary. But in geometry and dynamics the position is different. We are not at liberty to use *any* definition we choose of such terms as 'line' and 'force': our hands are to some extent tied. True: the axioms of geometry and the laws of motion, taken alone, establish the framework for drawing a picture of the world, and do not tell us directly what particular natural processes occur in that world. To that extent, they derive their necessity from a similar origin as truths of definition. Yet the framework they define is *not* an arbitrary one: it is one whose use can lead to coherent results, rather than to confused ones. This fact makes all the difference.

In science, our aim is to draw coherent intellectual pictures of Nature, and this demand limits our range of choice when it comes to defining our fundamental terms. Elsewhere, perhaps, definitions can be arbitrary. But the scientist's ideas gain precision as his knowledge of Nature increases: the words he uses have to be re-defined in the light of his discoveries. In this joint progress, his definitions become less and less arbitrary. If his theory is to be successful, the form of his ideas must be closely tailored to the natural world.

In retrospect, we can see that the crucial watershed in a science is reached when men hammer out a system of ideas that provides a coherent and comprehensive picture. Up to that point, there is groping and uncertainty; definitions of crucial terms remain unclear; ground is won only to be lost again; and the progress of the science is not cumulative.

At the opening of our present story, many distinctions were still obscure, which the growth of astronomy and dynamics had to clarify and establish. As men's ideas about the Heavens and the Earth developed, so they repeatedly changed their beliefs, their conceptions—and even the very hopes and ambitions behind their search for the System of the World and the Mechanism of Heaven and Earth.

At first they looked only at the raw material of astronomy: the changing appearances of the sky. Led by their critical curiosity, they next pursued familiar analogies to explain the things they saw going on above their heads. Later still, they conceived the possibility of fathoming the mathematical layout of the heavens, and even the forces at work in it. At times, they were doubtful of their right to aspire so high, and fell back out of modesty on to bare mathematical analysis of appearances. Partly, these doubts sprang from an inadequate grasp of the physics of the heavens, partly they sprang from a desire not to trespass into theology. Aristotle's first attempt at a synthesis was premature, for lack of an adequate system of mechanics. But a better mechanics was eventually built up, on the foundations laid by the mediaeval scholastics. The second assault on dynamics and planetary theory ended in success. Between 1540 and 1690 the last relics of the older synthesis were displaced by a new picture, which has remained the foundation of our 'common sense' right up to the present day.

In this story we have watched men studying the universe, as it were, through an intellectual microscope. Where, to begin with, they could make out only blurred outlines in ambiguous proportions, now their picture is crisp and sharp. This change has been accomplished not just as a result of honest observation and reporting. Of equal importance has been the progressive re-shaping and re-focussing of the fundamental ideas which form our intellectual instruments.

FURTHER READING AND REFERENCES

The immediate reception of Newton's theory is documented in

H. G. Alexander (editor): *The Leibniz-Clarke Correspondence*

The wider influence of Newton's ideas, in science, literature, and general thought, is discussed in

I. B. Cohen: *Franklin and Newton*
Basil Willey: *The Eighteenth-Century Background*
Marjorie Nicolson: *Newton Demands the Muse*
A. Wolf (and D. McKie): *A History of Science, Technology and Philosophy . . . in the Eighteenth Century*

Voltaire's *Lettres sur les Anglais* are well worth reading, and his story *Candide* includes a good deal of satirical reflection on the intellectual debates discussed here: the attitudes which he is criticizing are interestingly documented in the book

Sir Thomas Kendrick: *The Lisbon Earthquake*

On the general development of physical and astronomical ideas leading up to the theory of relativity, see

C. C. Gillispie: *The Edge of Objectivity*
A. Einstein and L. Infeld: *The Evolution of Physics*
W. K. Clifford: *The Common Sense of the Exact Sciences*
H. Poincaré: *Science and Hypothesis*

Theories of action at a distance are discussed in a monograph by Mary B. Hesse, and several of the philosophical points raised in this last chapter are carried further by N. R. Hanson in his book *Patterns of Discovery*. See also

P. Duhem: *The Aims and Structure of Physical Theory*

Index

Academy (Athens), 87, 93
  motto over entrance of, 79
  suppressed, 148, 153
Accademia Dei Lincei, 184
Acceleration
  Aristotle's lack of understanding of, 101-103, 211, 213
  Beeckman on, 217
  formula for, 216
  Galileo on, 21, 100, 218-220
  Jordanus of Nemours on, 214
  in Merton School, 214-215
  Newton on, 101, 211, 217
  Oresme on, 215-217
  Strato on, 133, 214
Acoustics, Pythagorean, 71
Acropolita, 154
Adams, Couch, predicts Neptune, 255
Air, *see Pneuma*
Alchemy
  Kepler's theories and, 207
  Newton and, 243
Alembert, Jean d', on conservation of motion, 253
Alexander, H. G., *The Leibniz-Clarke Correspondence,* 271
Alexander of Aphrodisias, on Pythagoreans, 72-73
Alexander the Great, 40, 131-132
Alexandria, 58, 130, 153, 156
  astrology in, 145
  astronomy in, 40
  burning of library in, 148
  as scientific centre, 128, 132-133
Alfonsine Tables, 161-162
Alfonso X, King, 162
Al-Mansur, Caliph, 156
Anaxagoras, 72
  astronomy and meteorology of, 57, 68-69, 108
  biography of, 68, 126
Anaximander, astronomy and meteorology of, 66-67
Anaximenes, astronomy and meteorology of, 67-68
Andrade, E. N. da C., 249
Angular measure, Babylonian origin of, 26
Apeiron, meaning of, 65-66
Apollonios, epicyclic theory of, 121-122, 138
Aquinas, Thomas
  cosmology of, 162
  harmonizes Aristotle with theology, 160-161, 183, 267-268
Arabs; *see also* Islam
  on projectiles, 221
  translate Greek sources, 53, 155
Archimedes, 58, 119, 132
  on Aristarchos' theory, 122-123
  'centre of gravity' in, 134
  kinematics of, 213
  'limits' in, 134, 151
  re-published in Venice, 183
  squares the circle, 149-151
Aristarchos of Samos, 89, 164, 165, 197
  heliocentric theory of, 116, 122-127
Aristophanes, *The Clouds,* 57
Aristotle, 53, 54, 89, 213, 248
  Alexander the Great and, 131
  Aquinas and, 160-161, 183, 267-268
  cosmology of, 69, 96, 105-112, 145
    Greek criticism of, 115-116, 119-127
  dynamics of, 92-104
    Philoponos' commentary on, 117-118
  on force and resistance, 225
  heavenly bodies as animate in, 109-110
  mechanics in, 211-212
  'Mechanics', 133
  Metaphysics, 160
  Newton on, 233
  On Physics, 63, 92, 104, 213
    Averroes' commentary on, 160
    Philoponos' commentary on, 117-118
  On the Heavens, 72, 92
  Physical Problems, 120
  Plato and, 93
  Ptolemy on, 143
  on Pythagoreans, 72, 73
  on Thales, 65
  'Unmoved Mover' in, 106, 160, 179
  zoology of, 94, 109-110
Arithmetic; *see* Mathematics, Pythagoreans
Armenia, 155
Armillary sphere, *Plate* 4

Armitage, A., *The World of Copernicus* (*Sun, Stand Thou Still*), 181
Astrolabes, 156, 158
Astrology, 44
  astronomy and, 18
  Babylonian horoscopes in, 25
  Kepler and, 45, 147
  Newton and, 147, 239, 247
  Ptolemy's defence of, 146-147
  recovers ground in Greece, 130, 145
Astronomy
  confused with meteorology, 17-18
  cosmology and
    Osiander on, 177-178, 196
    separated in Middle Ages, 162
    united by Greeks, 48
  early problems of, 15-18
  instruments in; *see also* Telescope
    Arab, 156, 158
    Babylonian, 40
    Greek, 156
    Tycho Brahe's, 189, *Plate* 11
  Mediaeval stagnation in, 153-155
  physical vs. mathematical
    in Aristotle and Ptolemy, 128-129, 135-136, 139-144
    in Copernicus, 179-180, 197
    in Geminos, 135
    and Newton, 239
  Plato on, 81-84
Athens, 156
  academies of, 130; *see also* Academy; Lyceum
  calendar of, 31, 32, 34
  Golden Age in, 63
Atomic theory
  in Greek science, 58-59
  in Newton, 265
Averroes, commentary on Aristotle of, 160
Axioms, as true principles of Nature, 70

Babylonians; *see also* Mesopotamia
  archives of, 24-25
  astronomical records of
    accuracy of, 33-34, 39-40, 41
    clay tablets as, 25, *Plate* 1
    Greek use of, 24-26, 40, 41, 48, 136-137
  astronomy of, 61, 91, 136
    celestial forecasting in, 27-28, 30-35, 37-40
Babylonians, astronomy of (*cont.*)
    computation of conjunctions in, 48-50
    theory of Moon's phases in, 45-47
  bibliography on, 50-51
  calendar of, 30-34
  creation myth of, 42-44
  geometry of, 26, 70
  interconnection of religion and civil affairs of, 45
  numerals of, 26
Bacon, Francis, 184, 230
Baghdad, as scientific centre, 155-156
Barrow, Isaac, 229, 230
Bayeux tapestry, *Plate* 7
Beeckman, Isaac, on acceleration, 217
Benedetti (physicist), on impetus, 223, 226
Bentley, Richard, 110
  on gravity, 256-257
Berossos, lectures on astronomy, 45-47, 65
Bessel (astronomer), 261
Blondell, 266
Blundeville, Thomas, on Copernicus, 176
Boethius, 159
Book of Hours (Duc de Berry), *Plate* 8
Borelli, (astronomer), on attraction of Sun, 227, 233, 239
Bradwardine, Thomas, 225, 227
Brahe, Tycho, 125, 178, 209
  cosmology of, 184-187, 188-189, *Plate* 12
  Kepler and, 202-203
  observes comet, 188-189
  observes super-nova, 187-188
Breasted, J. H., *Ancient Times*, 50
Brodetsky, S., 249
Browning, Robert, *The Bishop Orders his Tomb*, 158
Bruno, Giordano
  cosmology of, 190-191, 213, 225, 237
  executed, 191
Bunyan, John, *Pilgrim's Progress*, 84
Buridan, John, 'impetus' of, 221-223, 239
Burnet, J., *Early Greek Philosophy*, 89, 114
Burtt, E. A., *Metaphysical Foundations of Modern Science*, 249
Butterfield, H., *The Origins of Modern Science*, 181, 249
Byzantium; *see* Constantinople

Cairo, 53
Cajori, F., 249
Calculus
  Archimedes' anticipation of, 134, 151
  infinitesmal, 105, 134, 151, 214
  invention of, 104-105
    controversy over, 256
  of Leibniz, 151, 256
  of Newton, 151, 211, 241, 248, 256
  Oresme's anticipation of, 215-217
Calendars; *see also* Month; Year
  'leap-days' in, 33, 137
  lunar and solar, 31-34
Callippos
  cosmology of, 107
  Newton on, 233
  uses Babylonian data, 136
Callisthenes, 131, 136
Caspar, Max, *Johann Kepler,* 209
Cassiopea, 187
Cause-and-effect
  establishment of ideas of, 18
  among Stoics, 146
Celestial forecasting; *see* Forecasting, astronomical
Chaldeans; *see* Babylonians
Change, 19
  in Aristotle's thought, 94-95
  first recognized in sphere of fixed stars, 188-189
Childe, V. Gordon, *What Happened in History,* 50
China
  Galileo discussed in, 196
  super-nova recorded in, 188
Christianity, 109; *see also* Aquinas, Thomas; Protestantism
  Copernican system and, 177, 191, 197
  incorporates Aristotle, 115
  and mediaeval revival of science, 160-161, 165
  opposes Greek philosophy, 133, 148-149, 154
Clagett, M.
  Greek Science in Antiquity, 114, 152
  The Science of Mechanics in the Middle Ages, 180, 227
Clarke, Samuel, 256, 271
Cleanthes, 126
Clifford, W. K., 260
  The Common Sense of the Exact Sciences, 272
Cohen, I. B., 249
  Franklin and Newton, 272
Comets, 17
  in Bayeux tapestry, *Plate* 7
  Newton on, 233, 238
  no longer considered as portents, 21, 91, 239
  prediction of, 239, 247
  Tycho Brahe's, 188-189
Common sense
  Newton's theory becomes, 245, 250
  science and, 15-17, 47, 91
Comnena, Anna, 154
Conjunctions, Babylonian computation of, 48-50
Constantinople, 133, 153-154
  bibliography on, 180
  libraries of, 53, 154
  sack of, 158
Cooper, Lane, *Aristotle, Galileo, and the Tower of Pisa,* 227
Copernicus, Nicolaus, 74, 128, 162, 182
  as 'ancient' astronomer, 179
  bibliography on, 180-181
  Commentariolus, 170, 175
  cosmology of, 163-165, 169-180, *Plate* 10
  criticizes Ptolemy, 128, 140, 169-170, 178
  heliocentric theory of, 122
    basic assumptions of, 171-172
    criticized by Donne, 186, 198
    difficulties of, 125, 126, 173-179
    as reinstatement of Aristotle, 163-165
  On the Revolution of the Heavenly Orbs, 164, 173, 175
    placed on Index, 161
  so-called 'Revolution' of, 164
  on stellar parallax, 172, 173, 185
Cordova, library in, 53, 158-159
Cornford, F. M.
  From Religion to Philosophy, 89
  Plato's Cosmology, 89
  The Laws of Motion in Ancient Thought, 114
Cos, 45
Cosmas, mocks Greek astronomers, 154
cosmology; *see also* Astronomy; Babylonians
  Aristotle's, 69, 96, 105-112, 145

cosmology (*cont.*)
of Aristotle's successors, 115-127
arithmetical vs. geometrical, 79
Bruno's, 190-191
Copernican, 163-165, 169-180, *Plate* 10
Digges', 190-191, 194, 213, 225
18th century, *Plate* 13
of first systematic observers, 29-30
Galileo's, 198
'ingredients' vs. 'axioms' in, 69-71
Kepler's, 199-208, 247
mediaeval, 160-161, 162
modern, 261-264, 267-268
as myth, 63
Newton's, 233-249
Plato's, 84-89, 115, 120
pre-Socratic, 65-69
Ptolemy's, *Plate* 9
Pythagorean, 71-74, 79
Tycho Brahe's, 184-187, 188-189, *Plate* 12
Xenophanes', 56
Crab (constellation), 188
Creation, 223; *see also* God
in Anaxagoras, 69
in Anaximander, 66
in Anaximenes, 67
in Aristotle, 69, 268
in Babylonian myth, 42-44
continual, 263
'nebular' hypothesis and, 252, 261-264
in Newton, 251, 252, 266
in Plato, 63, 266
Crombie, A. C., *Mediaeval and Early Modern Science,* 180
Crusades, 158
Cusa, Nicolas of, on space and motion, 165, 168-169, 245, 262
Cycles, early observation of, 18

Dante, cosmology of, 162-163
Demokritos, 28
De Santillana, G., *The Trial of Galileo,* 197, 209
Descartes, René, 236
straight-line motion in, 226, 231, 239
vortex theory of, 231, 233
refuted by Newton, 241-242
Digges, Thomas, cosmology of, 190-191, 194, 213, 225
Divination
Babylonian, 31, 44
Stoic, 146
Donne, John, 186, 196
on cosmology, 198
Drake, Stillman, *Discoveries and Opinions of Galileo,* 209
Dreyer, J. L. E., *A History of Astronomy from Thales to Kepler,* 89, 114
Duhen, P., *The Aims and Structure of Physical Theory,* 272
Dunkirk, 'miracle' of, 266
Durrell, Lawrence, *Bitter Lemons,* 23
Dynamics; *see also* Acceleration; Force; Hydrodynamics; Mechanics; Resistance
in Aristotle, 92-104
astronomy and, 15
defined, 19
as mathematical theory, 19-20
planetary, 105, 208; *see also* Planets

Earth
as cylindrical
in Anaximander, 66
as flat
in Anaxagoras, 68
in Anaximenes, 67
rotation of
in Copernicus, 176
Herakleides on, 120-121, 122
Nicolas of Oresme on, 165-168
Ptolemy's rejection of, 126
size of
Greek calculation of, 112-114
as sphere
in Aristotle, 110-112
concept mocked by Christians, 154
in Pythagorean thought, 72
rejected by Lucretius, 146
translation of
rejected by Greeks, 122, 125-126
Earth and Sky (film), 51, 152, 181, 209
Earthquakes, 17
Babylonian records of, 25, 38
Lisbon, 266-267
Eccentrics
Copernican, 175
in Greek astronomy, 138
Eclipses, 66; *see also* Moon; Sun
Education, in Plato, 81-84

Egypt
calendar in, 33
mathematics in, 70
Einstein, Albert, 243
The Evolution of Physics, 272
theory of relativity of, 250, 255-256, 259-260, 264
Enuma Elish, 42-44
Ephemerides, 41, 47
defined, 25
Epicurus
opposed to astronomy, 146
scientific attitude of, 60
Epicycles
in Apollonios, 121-122
in Copernicus, 175, 176
illustrated, 141-142
in Ptolemy, 138-140
Equants
criticized by Copernicus, 140, 170-171, 175
in Ptolemy, 138-139
Equinoxes, *see* Precession of the equinoxes
Eratosthenes, estimates circumference of Earth, 112-113
'Ether', 257, 265
Euclid, 26, 58, 80, 159
Eudoxos of Knidos, 80, 136, 213
cosmology of, 87-89
Aristotle's use of, 105-107
Newton on, 233
Experimental method
Galileo's 'invention' of, 220-221
why Greeks did not adopt, 61
Extrapolation, 144
in forecasting tides, 36

Farrington, B., *Science in Antiquity,* 89
Festugière, A. J.
Epicurus and his Gods, 152
La Révélation d'Hermès Trismegiste, 152
Flamsteed, John, 236
Force; *see also* Resistance
in Aristotle, 97-98
centrifugal, 226
centripetal, 234-236
'fields' of, 258-260
mathematical theory of, 81
mediaeval view of, 225
Forecasting
astronomical
Babylonian, 27-28, 30-35, 37-40
rests on empirical techniques, 247
of tides, 35-37, 247
Frankfort, H., *Before Philosophy,* 51
Frauenhofer, Joseph von, spectrum analysis of, 262
Fulbert, Bishop, 157-158, 159

Gade, J. A., *The Life and Times of Tycho Brahe,* 209
Galen, 54, 61
Galileo Galilei, 19, 93, 178, 184, 236
on acceleration, 21, 100, 218-220
bibliographies on, 209, 227
cosmology of, 198
Discourses on Two New Sciences, 197, 219
impetus in, 224-225
'Leaning Tower experiment' of, 100, 117
on mathematics, 249
mechanics in, 218-221
on 'nebulous' stars, 261
on rotation of Earth, 167
scientific interests of, 189-190
Sidereus Nuncius, 191, 196
telescopic observations of, 186, 191-196, 197
The Two Chief Systems of the World, 197
on tides, 237
trial of, 197
Gamow, George, 168
Geminos, on physical and mathematical astronomy, 135
Geocentric theory of planetary motion, *Plate* 9; *see also* Heliocentric theory of planetary motion; Ptolemy
as sophisticated theory, 16-17
Tycho Brahe's, 185-187, *Plate* 12
Geometry
Babylonian, 26, 70
Greek, 59; *see also* Archimedes
Euclidean, 26, 58, 80
Platonic, 79-82, 206
Pythagorean, 70, 74-79
Gerard of Cremona, 157, 161
Gilbert, William, 206

Gillispie, C. C., *The Edge of Objectivity,* 227, 272
God
as Divine Architect, 183, 266-267, 268-269
as geometer, 206
in mediaeval cosmology, 160, 162
Newton's views on, 110, 251, 252-254
as *Nous,* 69
as Unmoved Mover
in Aristotle, 106, 179
in mediaeval cosmology, 160
Gods
heavenly bodies as
in Aristotle, 108-109
among Babylonians, 25, 41-45
in magic and science, 62-63
Graphs, for acceleration, 214, 215-217
Gravitation
in Archimedes, 134
in Buridan, 222
in Einstein, 255-256, 259-260, 264
in Kepler, 206
in Newton, 134, 208, 232, 235-238
difficulties arising from, 256-260
Greece; *see also* Athens; Cosmology; Geometry, Greek
astronomy in
Babylonian influence on, 24-26, 40, 41, 48, 136-137
Newton on, 233
bibliographies on, 89, 114, 152
contrasted with Mesopotamia, 54-56
kinematics in, 213-214
science in; *see also* Physics, Greek
basic principles of, 58-64
condemned by Christians, 133, 148-149, 154
decline of, 128-130, 144-149
development of theory in, 64-79
mediaeval attitude toward, 159-161
sources of, 52-54
unpopularity of, 57-58
Greek manuscripts, snob value of, 158
Guthrie, W. K. C., *The Greeks and their Gods,* 89

Hail, 17
Anaximenes on, 67
Hall, A. R., *The Scientific Revolution,* 209, 249
Halley, Edmund
comet of, 239, 247
Newton and, 21, 230, 243-244
Hammurabi, 42
sets up calendar, 31-32
Hansen (astronomer), 39
Hanson, N. R., *Patterns of Discovery,* 272
Harmony of the planets
in Kepler, 206-207
Pythagorean, 73
Heath, T. L.
Aristarchus of Samos, 89, 152
Mathematics in Aristotle, 114
Heliocentric theory of planetary motion, 16-17; *see also* Copernicus; Galileo; Kepler; Newton
of Aristarchos, 116, 122-127
Greek objections to, 123-127
as possible Pythagorean theory, 72-74
Helium, discovered, 263
Heraclitus, 268
Herakleides of Pontos, 165
cosmology of, 120-121
Herschel, Sir John
discovers Uranus, 254
Hesiod, Xenophanes on, 56
Hesse, Mary B., 272
Heytesbury, William, 227
on acceleration, 214-215, 220
Hipparchos of Rhodes, 117, 132
discovers precession of the equinoxes, 137
loss of work of, 53, 134
uses Babylonian records, 40, 41, 137
uses eccentric, 138
Homer, Xenophanes on, 56
Hooke, Robert, 240
Newton and, 230, 233, 239
Horrocks, Jeremiah, 229
Hoyle, Fred, 168
Hubble, Edwin P., 'red-shift' of, 263
Hume, David, 265
Huygens, Christiaan, 239, 240
on centrifugal force, 226-227
Hydrodynamics
Newton and, 241
Stokes' Law in, 99
Hydrogen, continual coming-into-existence of, 263

Impetus; *see also* Inertia; Projectiles
Benedetti on, 223
Buridan on, 221-223
India, astronomy in, 156
Inertia, Newtonian, 224
Inquisition, 160, 161
Instruments, *see* Astronomy; Telescope
Intercalations, 33-34
Ionians, 53, 55, 56, 91, 268; *see also* Anaximander, Anaximenes
Iraq Petroleum Company, 51
Irene, Empress, 154
Irrational numbers, 79
Islam, 180; *see also* Arabs
translations from Greek in, 53, 155, 156
scholarship of, 155-158
Italy, 53, 55, 184

John of Holywood, *De Sphaero,* 162
Jordanus of Nemours, 214
Jundishapur, 155, 156
Jupiter, 201
density of, 207
motion of, 207
in Vitruvius, 29-30
in *Myth of Er,* 85-86
satellites of
discovered by Galileo, 195-196
Kepler's Third Law and, 236
Justinian, Emperor, 148

Kant, Immanuel
on gravitation, 257
nebular hypothesis of, 168, 251, 261
on science, 269-270
Kendrick, Sir Thomas, *The Lisbon Earthquake,* 272
Kepler, Johannes, 164, 178, 209, 213, 248
aims of, 198-199
as astrologer, 45, 147
cosmology of, 199-208, 247
Mysterium Cosmographicum, 199
New Astronomy, 203
Newton and, 105, 233
Platonic solids in, 80, 200-201, 205-206
Pythagoreans and, 73, 74
stellar parallax and, 185
Tycho Brahe and, 202-203
Kepler's Laws of Planetary Motion, 199
Newton's use of, 231-232
Second Law, 203, 206
illustrated, 204
in Newton, 235-236
Third Law, 207-208
Huygens on, 227
in Newton, 236, 242
Keynes, Lord, on Newton, 243
Kidinnu (astronomer), 39-40
Kinematics; *see also* Mechanics
defined, 213
planetary
defined, 105
Kirk, G. S. and Raven, J., *Stoics and Epicureans,* 152
Koestler, Arthur, *The Sleepwalkers,* 209
Koyré, A., 249
Etudes Galiléennes, 227
From the Closed World to the Infinite Universe, 227
Kuhn, T. S., *The Copernican Revolution,* 114, 180

Lactantius, dismisses idea that Earth a sphere, 154
Laplace, Marquis de, 251, 261
planetary calculations of, 254
as social scientist, 265
Lavoisier, Antoine, 264
'leap-days'
in Egyptian calendar, 33
skipped every 400 years, 137
Leibniz, Baron von, 267, 271
on acceleration, 217
calculus of, 151, 256
rejects Newton, 253, 256, 258
Leverrier, Urbain
predicts Neptune, 254
predicts Vulcan, 255
Libraries
Alexandria, 132
burned, 148
Baghdad, 53
Cairo, 53
Constantinople, 53
Cordova, 53, 158-159
Paris, 159
Light, Newton's theories of, 265
Lisbon Earthquake, 266-267
Locusts, in Babylonian records, 25, 38

Lucretius, dismisses sphericity of Earth, 146
Lyceum (Athens)
established by Aristotle, 93
successors of Aristotle at, 133

Magic, concept of order as basis of, 62
Marduk, 63
in *Enuma Elish,* 42-44
Mars, 201
density of, 207
motion of
in Apollonios, 122
Babylonian forecasting of, 27-28
diagram of, 27
in Kepler, 202-205
in Sosigenes, 119
in Vitruvius, 29-30
in *Myth of Er,* 85-86
Tycho Brahe's observations of, 185
Mathematics; *see also* Calculus; Geometry; Numbers
abstract
invention of, 70-71
in Aristotle, 94
Babylonian, 26
as basis of theory, 70-71, 80
as bridge between physics and theology, 143-145
dynamics as branch of, 19
theory of force and, 81
Matter
in Aristotle, 108
in Plato, 80
Maxwell, James Clerk, 91
Mechanics; *see also* Dynamics; Statics
Aristotelian vs. Newtonian, 211-213
definitions in, 211, 217-218
Galileo and, 218-221
in Middle Ages, 210
Mercury, 201
density of, 207
motion of
in Apollonios, 121-122
in Herakleides, 120
irregularities in, 255
in Vitruvius, 29-30
in *Myth of Er,* 85-86
Merton College (Oxford), 214-215, 218, 219
Mesopotamia, *see also* Babylonians
stability of life in, 54-55
star-worship in, *Plate* 5
Meteorology
origin of term, 17-18
in pre-Socratic science, 65-69
Middle Ages
astronomical observations in, 202
mechanics in, 210
science in
bibliographies on, 180-181, 209, 227
European, 153-155, 157-180, 182-184
Islamic, 155-158
Mieli, A., *La Science Arabe,* 180
Miletus, *see* Ionians
Milky Way, 17, 261, *Plate* 14
Anaxagoras on, 68
Galileo on, 193-194, 261
Milton, John, 196
Molière, 257
Momentum
development of concept of, 102-103
Philoponos on, 117-118
kinetic energy and, 253
Months, lunar, 31-33
Hipparchos' estimate of length of, 137
Moon; *see also* Tides
in Aristotle, 232
distance from Earth of
Ptolemy's calculations of, 113-114
early attitudes toward, 18
eclipses of
in Anaxagoras, 65, 68
in Aristotelian School, 111, 120
Babylonian forecasting of, 25, 30, 31, 38-39
in Eudoxos, 87
Galileo's observations of, 191-192
motion of, 28
Newton on, 232, 240-241
in Vitruvius, 29
in *Myth of Er,* 85-86
phases of, 33
in Anaximander, 66
Bernossos' theory of, 45-47
New Moon, 30-32
tides and, 37
rotation around Earth of
Galileo's defence of, 194-196
Mongol invasion, 55
More, L. T., 249

Motion; *see also* Impetus; Mechanics; Planets
Aristotle's theory of, 95-104, 248
conservation of, 252-254
straight-line vs. circular, 223-227
Music
Kepler and, 206
Pythagoreans and, 71
Myth of Er, 84-86, *Plate* 4
Myths
Babylonian, 42-44, 63
denounced by Xenophanes, 56
Greek, 63

Nature, *see* Cosmology
Nautical Almanac Office, *Astronomical Ephemeris,* 25, 41, *Plate* 1
Nebuchadnezzar II, 34
Nebular hypothesis, 252, 261-263
Neo-Platonists, 144-145, 174
Neptune, discovered, 254-255
Neugebauer, O., *The Exact Sciences in Antiquity,* 50, 152
Newton, Sir Isaac, 18, 19, 93, 126, 164
on acceleration, 101, 211, 217
astrology and, 147, 239, 247
Biblical scholarship of, 243
bibliographies on, 249, 271-272
biography of, 229-230, 243
calculus of, 151, 211, 241, 248, 256
chemistry of, 243
on comets, 21, 91, 233, 238, 239
cosmology of, 233-249
general estimate of, 228-229
on God and creation, 110, 251, 252-253, 266
on gravitation, 208, 235-238
apple story and, 232
difficulties arising from, 256-260
inertia in, 224
Kepler and, 105, 233
laws of motion of, 99, 101, 103, 269
Leibniz and, 253, 256, 258
on light, 265
mechanics in, 211-212
Opticks, 241, 243
Philosophiae Naturalis Principia Mathematica, 236, 243
Halley and, 21, 230, 243-244
The System of the World, 232
wider influence of, 264-268
Nicolas of Cusa; *see* Cusa
Nicolas of Oresme; *see* Oresme
Nicolson, Marjorie
Newton Demands the Muse, 272
Science and Imagination, 209
Nile River, 68
Nous, in Anaxagoras, 69
Numbers
irrational, 79
Pythagorean emphasis on, 74-79

Observations, astronomical, in Middle Ages, 202
Olmstead, A. T., *History of the Persian Empire,* 50
Oppolzer (astronomer), 39
Oresme, Nicolas (Nicholas) of
graphs acceleration, 215-217, 220
on impetus, 221, 224
on rotation of Earth, 165-169, 197
on temperature, 217
on velocity, 218
Orion, depicted by Galileo, 192, *Plate 6d*
Orreries, *Plate* 13
Osiander, 268
on limitations of Copernicus' work, 177-178

Parmenides, 70
Pasternak, Boris, 197
Pericles, 57
Persia, scholarship in, 148, 155
Philip of Macedonia, 94
Philoponos, 134
becomes Christian, 148, 149
on force and resistance, 225
on Momentum, 117-118, 221
Philosophy; *see also* Science
opposed by religion, 133, 148-149
to promote piety, 147
Physics; *see also* Dynamics; Mechanics; Statics
Greek
anaesthetized by Christianity, 149
in astronomy, 88-89, 135-136, 140-145
modern criticism of, 63, 92-93
mathematical; *see also* Astronomy, physical vs. mathematical
based on Plato, 80
Pirenne, H., *Economic and Social History of the Middle Ages,* 180

Planets; *see also individual planets*
density of
in Kepler, 206-207
distance of
Aristotle and Sosigenes on, 119-120
'harmony' of, 73, 206-207
movement of; *see also* Kepler's Laws of Planetary Motion
in Anaximenes, 67
in Aristotle, 96, 105-108, 233
in Buridan, 223
in Newton, 233-238
in Plato and Eudoxos, 85-87, 233
in Pythagorean thought, 72
perturbations in orbits of, 252-254
predicted from Newton's theory, 254-255
retrograde motion of; *see also* Epicycles
in Copernicus, 172-173, 179
in Eudoxos, 80, 87-89
illustrated, 27-28
tables for, 161-162
in Vitruvius, 29-30
Plato, 53
Aristotle and, 93
astronomy in, 81-84
cosmology of, 84-89, 115, 120
on creation, 63, 266
geometry of, 79-82, 206
Renaissance interest in, 212
Republic, 81-83, 84
on science, 144
theory of matter of, 80
Timaeus, 63
Platonic solids
Kepler's work with, 200-201, 205-206
Theaetetus' theorem on, 80, 200
Pledge, H. T., *Science since 1500,* 249
Pleiades, 193
Pluto, 254-255
Pneuma (Air)
Anaxagoras on, 68
Anaximenes on, 67
Poincaré, H., *Science and Hypothesis,* 272
Precession of the equinoxes
in Copernicus, 172
discovered by Hipparchos, 137
Printing, effect on science of, 183-184
Projectiles
Arabs on, 221
Projectiles (*cont.*)
Aristotle on, 102-103, 221, 222
Buridan on, 221-222
Philoponos on, 118, 221
Protestantism, and heliocentric theory, 177, 183, 250
Prussia, Duke of, 162
Ptolemy, Claudius, 54, 88, 132, 152, 154, 213
Almagest, 53, 146, 152, 156, 179
Arabic source of title of, 137-138
'rewritten' by Copernicus, 175
on astrology, 146-147
calculations of Sun's and Moon's distances by, 113-114
cosmology of, *Plate* 9
criticized by Copernicus, 128, 140, 169-170, 178
on distance of stars, 124
epicycles of, 138-140, 152
mathematical methods of, 138-142
Tetrabiblos, 146
thinks physics irrelevant to astronomy, 135-136, 140-145
on translation and rotation of Earth, 125-126
uses Babylonian work, 40, 41
Ptolemy Soter, King, 132
Pythagoreans, 24, 91
cosmology of, 71-74
'harmony of the soul' of, 71, 109
mathematics of
mathematical figures in, 74-76
religious significance of, 58, 71
Theorem of, 70, 77-79
Pythagoreon, *Plate* 3

Quintessence', in Aristotle, 108, 262

Rainbows
Anaxagoras on, 68
Anaximenes on, 67-68
Xenophanes on, 56
Ralph of Liège, 159
Ramsay, Sir William, spectrum analysis of, 262-263
Reason
Copernicus' beliefs on, 178
among Greeks
as main theme in philosophy, 62-63, 70-71

Reason, among Greeks (*cont.*)
its efficacy later doubted, 130, 145, 147
in 17th century thought, 182-183
Red-shift,' in distant galaxies, 263
Reginbald of Cologne, 159
Reinhold, Erasmus, *Prutenic Tables,* 162
Relativity, 250, 255-256, 259-260, 264
bibliography on, 272
Religion; *see also* God; Philosophy; Theology
heliocentric theory and
in Christianity, 177, 191, 197
in Greece, 126
interconnected in Babylonia with civil affairs, 45
Pythagorean mathematics and, 58
Resistance; *see also* Force
in Aristotle, 98-101
mediaeval view of, 225
in Newton, 99
Rhodes, 132
Rivers of Time (film), 51
Rosen, E., *Three Copernican Treatises,* 180
Royal Society, 184
Newton and, 230
Swift on, 265
Runciman, S., *Byzantine Civilization,* 180
Russell, Bertrand, 70
Rutherford, Ernest, 89

Sambursky, S., 152
The Physical World of the Greeks, 89, 114
The Physics of the Stoics, 152
Samos, *Plate* 3
Satellites, artificial, conceived by Newton, 234-235
Saturn
motion of, 201, 207
in *Myth of Er,* 85-86
in Vitruvius, 29-30
Uranus and, 254
Science; *see also* Technology; Theory
academies of
in Europe, 184
in Greece, *see* Academy; Athens; Lyceum
common sense and, 15-17, 47, 91
craft vs. speculative traditions in, 246
Science (*cont.*)
development of
factors affecting, 129-133
Greek contribution to, 63-64
development of distinctions in, 16-19
effect of printing on, 183-184
as faith, 59
mediaeval revival of, 158-161, 182-184
observation and experimentation in, 61, 83-84, 90, 220-221
Plato on, 144
in politics and social affairs, 265-266
provisional nature of, 268-271
as union of theory and practice, 61-62
Shooting stars, 69
Sicily, 155, 157
Simplicius, 119, 148, 155
attacks Philoponos, 149
Singer, Charles, *From Magic to Science,* 180
Socrates; *see also* Plato
criticized for astronomical interests, 57
Sophocles, 174
Sosigenes, on distance of planets, 119-120
Southern, R. W., *The Making of the Middle Ages,* 180
Space
in Digges, 190
extent of, 168-169, 173, 262
Spain, 155, 157-159
Spectrum analysis, 262-263
Stades, length of, 113
Stars
catalogues of, 188
Babylonian, 26
of Hipparchos, 137
distance of; *see also* stellar parallax
in Greek theory 123-125
Galileo's observations of, 191-194
as outermost spherical shell
in Aristotle, 106-109
in Eudoxos, 85-87
in Tycho Brahe, 189
pre-Socratic explanations of, 65-69
spectrum analysis of, 262-263
spiral nebulae of, 261, *Plate* 14
worship of, 146, *Plate* 5
Statics, Archimedes on, 134
Stellar parallax
Copernicus on, 172, 173, 185
defined, 124
illustrated, Plate 6

Stellar parallax (*cont.*)
Kepler on, 185
proven in 19th century, 124, 194, 261
Ptolemy on, 124
Tycho Brahe on, 185, 186-187
Stevin, Simon, 117
Stoics, 145-146
Stokes' Law, 99
Strato, 119, 133
Strong, E. W., *Metaphysics and Procedures,* 249
Stukeley, William, 232
Struve, W. W., 261
Sullivan, J. W. N., 249
Sun; *see also* Heliocentric theory of planetary motion
in alchemy, 207
attraction of, 227, 233, 239, 240; *see also* Gravitation
Copernicus' respect for, 174
distance from Earth of
in Copernicus, 171
in Ptolemy, 114
eclipses of, 154
in Anaxagoras, 68
Babylonian forecasting of, 30, 31, 39
Einstein's theory 'confirmed' by, 256
in Eudoxos, 87
in Kepler, 199, 201, 206-208
motion of
in Newton, 237
in relation to Moon, 28
in Vitruvius, 29-30
in *Myth of Er,* 85-86
size of
in Anaxagoras, 68
spectrum analysis of, 262-263
Sundials, 28
Sunspots, Galileo's observations of, 196
Super-novas
of A.D. 1054, 188
Tycho Brahe's, 187-188
Swift, Jonathan, *Gulliver's Travels,* 265
Sylvester II, Pope, 157
Syracuse, 132
Syria, 155

Taylor, F. Sherwood, *Galileo and the Freedom of Thought,* 209
Technology
in late Greek era, 145
Plato on, 81-83
theory and, 61, 90-91
Telescope, Galileo's, 186, 191-196, 197
Temperature, Oresme on, 217; *see also* Thermometer
Tertullian, attacks scientific curiosity, 148
Tetraktys, 74, 76
Thales, 24, 65, 66
Theaetetus, theorem of, 80, 200
Theology, astronomy and, 251
Copernicus and Osiander on, 177-178
as independent disciplines, 20-21, 267-268
Ptolemy on, 143-145
in 17th century, 182-183
Theophrastos, 65-69
Theory; *see also* Science
analysis of events and, 39
concept of order as basis of, 62
craft and, 246
evolution of, 20-21
general
imagination and, 240
limitations of, 19
through union of branches of sciences, 91
invention of, 58-64
method of theoretical models in, 64-65
Ptolemy on, 139-145
Thermometer, Galileo's, 220; *see also* Temperature
Tides
calculation of, 35-37, 247
Newton's explanation of, 237-238
'Time-series', 35
Toledo, 157, 161-162
Tolstoy, Lev, on great men, 228-229
Townley (astronomer), 236
translations
of Arabic works into Latin, 157, 159
of Greek works by Arabs, 53, 155, 156
Tycho, *see* Brahe, Tycho

Universe, *see* Space
Ur, 25, *Plate* 2
Uraniborg, *Plate* 11
Uranus, 255
discovered, 254
Uruk, 25

Vacuum, impossibility of in Aristotle, 100-102
Velocity
instantaneous, 104, 248

Velocity (*cont.*)
lacking in Aristotle, 211, 213
Oresme on, 218
Venice, 158
re-publication of Archimedes in, 183
Venus, 201
density of, 207
movement of
in Apollonios, 121-122
in Herakleides, 120
in Sosigenes, 119
in Vitruvius, 29
in *Myth of Er,* 85-86
phases of, 173, 177
Vitruvius
on Bernossos' theory, 45-47
on motions of heavenly bodies, 28-30
Voltaire, 266
Candide, 267, 272
Lettres sur les Anglais, 272
Vulcan (supposed planet), 255

Willey, Basil, books of, 249, 272
Whitehead, Alfred North, on Universe as organism, 110
Wolf, A. and McKie, D., books of, 249, 272
Wooley, Leonard, *The Sumerians,* 50

Xenophanes, on anthropomorphism, 56

Years
Hipparchos' estimate of length of, 137
in lunar and solar calendars, 33
planetary
in Kepler, 207-208

Zeno of Citium, 145
Zeno of Elea, paradoxes of, 104-105
Ziggurat, 43-44, 45, *Plate* 2
Zodiac, 87
in Babylonian astronomy, 26
in Vitruvius, 29-30

on *The Architecture of Matter*

"[*The Architecture of Matter*] aims to retell the story of the evolution of scientific ideas from a fresh point of view. The authors review the various scientific ideas of animate and inanimate matter that were advanced from ancient times up to the present—the history, in other words, of physics, chemistry, and biology. . . . The book provides a frequently skillful, sometimes brilliant exposition of highlights, turning points and breakthroughs in the growth of scientific knowledge. . . . No other history of science is so consistently challenging."

—*Scientific American*

"*The Architecture of Matter* is to be warmly recommended. It is that rare achievement, a lively book which at the same time takes the fullest possible advantage of scholarly knowledge."

—Charles C. Gillespie, *New York Times Book Review*

"*The Architecture of Matter* is brilliantly successful. . . . At one stroke they have reintegrated science into the general intellectual tradition of the West, and removed its stultifying antithesis to the more "imaginative," "humanistic," and "discursive" modes of thought. Thus we now see the alchemists, for example, as giving an account of the elements not only scientifically valid in some ways but superior, *scientifically* superior, to that of later centuries."

—Martin Green, *The Observer*

on *The Discovery of Time*

When and how did scientists come to think of the world of nature as having been notably different in the past from what we experience in our own lifetimes? When and how did historians and theologians begin to consider the history of nature as relevant to their own ways of thinking? And in what way have our ideas about time developed over the centuries? Stephen Toulmin and June Goodfield argue that "In the whole history of thought no transformation in men's attitude to Nature . . . has been more profound than the change in perspective brought about by the discovery of the past."

Beginning with the early Middle Eastern chronologies, they trace the development of our ideas about the passage of time through Classical Greece and Medieval Europe, to Descartes, Vico, Kant, Herder, and Darwin.

"A discussion of [an] historical development of our ideas of time as they relate to nature, human nature, and human society. . . . [T]he excellence of [this book] is unquestionable."

—Martin Lebowitz, *Kenyon Review*

"A first-rate achievement of elementary historical synthesis."

—Thomas S. Kuhn, *American Historical Review*